Irresistible Attraction:

Secrets of Personal Magnetism

Kevin Hogan
Mary Lee LaBay

Irresistible Attraction:
Secrets of Personal Magnetism

Hogan, Kevin, 1961 –
LaBay, Mary Lee 1951 –
Irresistible Attraction: Secrets of Personal Magnetism

Bibliography
  1. Psychology
  2. Relationships
  3. Human Sexuality

ISBN 10: 0-9635085-2-0
ISBN 13: 978-0-9635085-2-2
CIP Data:
            302.12dc20   HM321.H67

by Kevin Hogan
© 2000, 2008
All Rights Reserved
First printed in USA
Also printed in Korea, Romania, Japan, Poland,
Thailand, Canada, Turkey, Nigeria, Indonesia

Co-author: Mary Lee LaBay

Illustrations: Jack Swaney

Published By:
Network 3000 Publishing
3432 Denmark Ave. #108
Eagan, MN 55123
Phone (612) 616-0732
Fax (800) 398-4642
Please Direct foreign rights inquiries to:
meta@ix.netcom.com

Table of Contents

Acknowledgements

Our very special thanks to Jack Swaney for his inspirational cartoon drawings placed throughout the book.

I (ML) would like to thank the many people who have given me support, insight and encouragement during the writing of this book. And to my family and friends who have been patient with my limited available time.

In the hopes of not leaving anyone out, I would like to give special appreciation to my parents, Maurice and Margery LaBay; my children, Quincy and London Miller; and to Trisha Nerny, Barry "Bear" VanderBrink, David Thusat, and Giavonne Mitchell, Debiruth Stanford, Nancy Parks, Rebecca Phifer, Pat Bauman. Special thanks to Lee Zehrer and Jennifer Fox at Kiss.com. And finally, to all the single men and women that helped in the research for this book.

I (KH) want to express my appreciation to Nancy Conger for developing the concept of Irresistible Attraction with me in 1997. I also want to offer a special thank you to Wendi Friesen for sharing with me her innermost thoughts about the thinking of women (both her ideas and those of some of her clients) about men. I also want to thank all the women who shared personal and intimate insights about how they perceive men.

Finally, I want to thank my wife, Katie, for allowing me to live while I talked with all of these wonderful women about the intimate and detailed experiences in their lives!

Introduction

Sex.
The need to affiliate.
Sex.
The desire for love.
Sex.
The desire to have a close-knit family.

The ways men and women think really do seem to differ in their short and long-term goals! We'll help you determine for yourself what specifically your desires and needs in a relationship are. This book is both informative and fun. It is thoroughly researched and yet easy to read.

We draw from the fields of social psychology, evolutionary psychology, hypnosis, neurolinguistic programming, and the personal experiences of many individuals who were kind enough to share their innermost thoughts with you. This book is designed to not only help you understand what is attractive to men and women, but to show you how specifically to become more attractive in the minds of those you want to influence.

Not much happens in an intimate relationship until two people first are attracted to each other. In the majority of attraction experiences, there is a component of sexual attraction. There is also often a component of emotional, intellectual or ideological attraction as well. All of these areas will be discussed in this book but we are going to primarily focus on personal magnetism, which is in large part, a synthesis of sexual energy and attraction.

One aspect of attraction that you may encounter is occasional frustration about what attracts us to each other. Don't worry! We all get frustrated when we aren't as ideal as we could be. There are no ideal people. There are only ideals. Read this book, enjoy, have fun and learn to radiate your charismatic energy to everyone around you.

ONE

Are You Irresistible?

In business, sales, social networking and even the singles scene, we all benefit in many life-changing ways if we enhance our level of attraction to those around us.

You may think it shallow that people respond to what may seem to be superficial qualities like our appearance or how much money we make, but superficial as it may be, the facts are undeniable. If people find you attractive in one or many ways, you will probably have a better job, get better treatment in most social situations, and you will probably make more money. People who are perceived as attractive, using various (subjective) criteria, are perceived to be happier, healthier and more emotionally stable by others around them.

People are more likely to be hired for jobs if they are perceived to be attractive by the interviewer. People are much more likely to be asked out for a second date if they are considered attractive. People are more likely to make more sales if they are thought to be attractive. People are more likely to want to be our friend if they think us attractive.

Years of scientific research have provided overwhelming evidence about the value of being socially and physically attractive. It really is true that people we are "magnetized to" have some set of charismatic characteristics that often seem indefinable. In this book we will not only define these characteristics, but we will also show you how you can develop the traits that make the most socially attractive people sparkle. In addition to this, you will learn what other people see when they see you, and what they hear when you talk. You will learn how people feel around you and you will learn how to easily take yourself to the next level of charismatic attraction.

These are just a few of the questions that will be answered in this book.

- What does "attractive" mean?
- Who says what "attractiveness" is?
- How can you change your level of attractiveness to everyone around you?
- Do you have to change who you are inside and become a "phony?"
- What importance are the qualities of who you are "inside" to how attractive you are?
- Is attraction important if I'm already married?
- Does it matter if I have charisma if I have a 9-5 job?

What are the actual benefits of being more attractive? Recent research indicates the following about the assumptions we all make about attractive people. We believe that attractive people are:

- more successful
- more intelligent
- better adjusted
- more socially skilled
- more interesting
- more poised
- more exciting
- more independent
- *more charismatic*
- more sexual
- have a happier marriage
- have more professional and social success
- have more fulfilling lives

These are assumptions that the world makes about attractive people. So imagine the heightened chances of having opportunities come your way, just because people assume these qualities in you, just by looking at you!

Scientific research reveals quite a bit about what we find attractive in other people. A portion of this research comes from studying the evolution of species and studying groups and societies from all over the world and not just our melting pot in the United States. Over millions of years, we have evolved to have certain predispositions to engage in certain behaviors and not others. The process of evolution is not interested in political movements or religion. Life simply evolves. The creationist acknowledges that evolution simply is a fact of life and the fact that life continues to evolve by no means rules out the place of a creator.

Men in almost all cultures have typically played the role of the provider, the hunter and the gatherer. Until technology advanced far enough in the 20th century, he had to do all of these things for himself and his family. It was his natural role. The reason is simple enough. For millions of years, when a man and a woman found each other attractive enough to mate, children resulted. Children are time-intensive and someone in a family needs to care for the offspring of the mating. That lifetime project has universally fallen to the woman.

The woman bares the brunt of the child-rearing responsibilities. She carries a baby for nine months and her ability to do other things becomes limited. Once the baby is born, nature has wired "Mom" to feed the baby. Research is clear that babies, on average, who are breast-fed are healthier and smarter than children who are not. Healthy genes are "nurtured" by those women who follow the evolutionary plan of caring for children. Moms have evolved to be pre-disposed to care for their children. This makes a woman far more vulnerable than a man. A woman's life can change immediately after having sex. One minute she is not pregnant, the next minute she could have a lifetime commitment ahead of her. That is part of what life on this planet is all about. Therefore, women tend to be a bit more careful about who and when they will engage in sexual relations with.

Women typically have about 20-35 potential childbearing years in their lives. As I look at my own family tree, I see that many of my ancestors back in the 16th-19th centuries had as many as 12-15 children. They needed those children to farm and be part of a family team to keep the family alive. Women are the glue that holds society together at the level of the family. In almost every species on this planet, the female is designed to give birth and when young need to be raised, it is almost universally the female that does so.

Evolutionary psychology suggests that women are hardwired to find certain traits and characteristics attractive in men. Those characteristics are going to be the keys to making a woman's life as comfortable as possible while she performs her evolutionary tasks of giving birth and caring for the children she will raise.

The rise in feminism in the middle and late 1900's hasn't changed what is attractive in a man to a woman's eye...not one bit. You can change the clothes and the day-job but you can't change the genetic hard drive that you are born with. Women for millennia to come will be attracted to the man who can (in 21st century terms) bring home the money, show love and kindness to the woman and her children and have the respect of his peers. 1000 years ago, women were attracted to a man who could provide plenty of food, a safe place to live to raise the children and someone who could fight off the enemies of the house, tribe or group.

The first filter most women use to evaluate a man at the unconscious level is his probable access to resources. (In the 21st century that means money, advanced education, ambition, work ethic, and status in the peer group.)

Little has changed in the last hundred, thousand or one million years.

Men, on the other hand, have a different set of evolutionary tendencies and problems. Men don't have the capacity to give birth. They aren't built to breast-feed the newborn children and all the sensitivity classes in the world will not make them the best choice to raise children when contrasted with women. Males are the pro-creators of the species. This evolutionary charge means that a man is designed to create offspring everywhere and anywhere he can. The survival of life on the planet is largely dependent upon men finding a healthy woman or women to populate the earth and pass along his genes from generation to generation.

Men are looking for what they have always looked for in a woman. Men at some deep unconscious level, want to have sex with a woman who is healthy. A sickly woman cannot care for the baby he is about to create. She would die off and leave the baby to the man, and that notion is contrary to his programming. Don't think for a minute that a man actually consciously thinks about all of this in the heat of passion. All of this internal communication takes place at a much deeper space in the mind body. In the heat of passion, the man simply wants to fulfill his inner desire to have sex. The woman, however, is aware that this moment of pleasure could be the beginning of a lifetime of nurturing a baby into childhood and on into adulthood. For a man, that thought is distant and nebulous.

Men are not dispassionate about children or the woman he wants to be with. On the contrary, men can be very loving, kind and caring to women and children. Some men can commit a lifetime of resources to his first family, though this is not necessarily the norm. Regardless, men are not just clean-shaven apes. They can develop other interests outside of sex, sometimes the interests even parallel that of the woman he is attracted to!

So who is this healthy woman that a man is attracted to? How did our ancestors know if a woman was going to be "healthy" or not? Before the advent of modern medicine, just a century or so ago, you would simply look at another person and determine as best you could if a person was healthy or not. Typically thin to "normal" people (though not to the point of starvation) live longer and in fact, they always have. Obese people, on average, die earlier. It's important to note that in some tribes, heavier people are considered healthier and this is probably because you can live longer in an obese body than a nutrition-starved body. Dying at age 45 in an obese body was better than dying at age 20 in a nutrition-starved body.

In the United States, starvation is rarely a problem. Obesity is a leading cause of early death. Thin people live significantly longer than the obese, and therefore make better mates on that evolutionary level.

Men are attracted to women who are "healthy"; also known as physically attractive. They are attracted to women who will not stray from the man. Men will share his resources with a woman, but he will not let that woman share his resources with another man. To ward off other males means the man has to use valuable time in fighting off another enemy. He does this all day in society (competitors in business that were once competitors and invaders for food and land) and he is not interested in warding off men who will come and take his mate. When jealousy is experienced, the evolutionary man will simply "kill" his competitor. In civilized society, this still happens every day, but much less so than it did thousands of years ago. Today there are other outlets for a man's jealousy, though the urge to violence is still primarily felt.

Men are not necessarily attracted to women with money or resources. Historically men had to pay a woman's family for the right to marry the woman. A man never expected money or a commodity to mate with a woman. In fact, it has been argued by many that men historically pay for almost all sexual relations with women. Putting that somewhat thought-provoking theory aside, there are numerous traits and characteristics in a woman that are important to a man, and we will discuss them in more detail later in this chapter.

> **The first unconscious filter a man has when evaluating woman is physical attractiveness.**

This book will focus on what is normally attractive and normally helpful in increasing our "face value" to others. There are plenty of exceptions to the rule, which is why the saying, "to each his own," continues to flourish. For now, we will look at the research about just what it is that we are seeking in other people. For the balance of this chapter, we will consider attraction in the context of interpersonal relationships that *could* become intimate or long-term.

The Universal Attractor

Before we delve into the deepest impulses and urges of men and women, we will look at the one great universal attractor.

> *Like Attracts Like*

Both men and women find people more attractive after they get to know other people. As people spend time together and learn more about each other, attractiveness increases. Being with people who share similar values, beliefs and lifestyles provides a greater sense of connectedness, and we rate people who are like us as better looking.

We also know from research that people with similar attitudes to their own are seen as more attractive. This can be an implicit understanding of their values, such as being seen in a church or organization where people share the same values. The more similar the values and attitudes, the more attractive people are to the person.

In one interesting study, the more often students saw a female confederate in their classroom, the more positively they rated her personality, even though no one in the room ever interacted with her! The two compelling arguments for this response are that the woman was perceived to be a student and thus like the other people in the room and second that we tend to like people the more often we see them.

This point of view is supported by research. One study revealed that the subjects determined that the people who live next door to them in an apartment building are more attractive, on average, than other people. Neighbors are more likely to interact with each other and become more familiar with them. A surprising finding was that people who live near the mailbox and entrances of an apartment building were also perceived as more attractive than average!

We will talk about this principle (the exposure effect) later in the book.

People also seem to be attracted to others who are about the same level of attractiveness. In other words, one "7" seems to attract another "7." Men tend to agree on what "7" means. Women tend to disagree. Here is how both sexes tend to evaluate each other.

What Do Women Find Irresistible in Men?

Women say they want to find a "good man" they can love. This is true to some degree. Women rate "love" as the most significant need they want fulfilled by a man. However, reality is different from our beliefs. Love is wonderful, but women seem to be attracted to other characteristics and traits long before love develops.

One recent study revealed that 22% of American women are still feminists but feminism hasn't slowed 5 billion years of evolution. What women want in the new millennium hasn't changed much from what women wanted millions of years ago! So, what is it that attracts a woman to a man?

Men at Work

There is no question that women respect and look for a man who works hard, more than just about any trait or characteristic. If a man is not afraid of hard work, she knows that she will always be cared for. A man who is willing to work long hours is seen as tenacious and someone who is reliable. Women know at some level that a man who works long hours is a person who is able to bring stability, especially in terms of financial reward. Women, on average, are attracted to men who work hard. Women don't normally consciously think, "Oh, good, he works 12 hours days, therefore I will have plenty of resources in my life." They simply see the man for the industrious worker that he is and intuitively know that this is the kind of a man who could supply her with what she will need to live a potentially happy, prosperous and safe life.

Of all the characteristics a man can have, this is one of the most important that women look for consciously and unconsciously.

> *If a man wants to be irresistible to women, he needs to subtly let them know that he works hard, often long hours, and that he isn't afraid to do so in the future.*

The Man Who Persists

Women often complain about men who continually call them on the phone. But be aware, women admire persistence in a man. Women like to see a man who can get up after he falls down and get back into the game. The ability to fail and then go back at it (whatever "it" is) is something women are drawn to. Women know that there will be tough times in life, and when they see someone who fails and is not deterred, they get a sparkle in their eye.

More marriages have been made than lost by a man who called a woman until she said, "yes" than you might imagine. The quality of persistence is something men need to develop if they are to be taken seriously as attractive. A man's persistence is a trait that is culturally admired by both men and women.

> *People who persist tend to succeed, and that is just one reason why women ultimately say, "yes" to persistent men.*

Men on Top

Women find men who have risen to "the top" most appealing. The top could be the CEO of a business, the pinnacle of their profession, the leader in a church organization or the top of any group. A man who is at or near the peak of the pyramid is someone who probably can give her the security and safety that she needs. Men at the top make more money than men on the bottom. Men at the top obviously have more power than men at the bottom of the pyramid. Power certainly is an aphrodisiac and money (resources) comes from the proper use of power. Money buys security, stability and allows for a more flexible lifestyle.

How important is money?

Almost all women surveyed in study after study indicate they want to be with a man who makes more money than they do. (Men on the other hand generally couldn't care less if a woman earns more than he does!) One woman recently said to me, "If he doesn't make more than I do, what is the point?"

Men at the top have the respect of their peers and this is important to women. Women feel good when the man they are interested in is not only able to provide them with financial rewards, but also a certain level of prestige. More important than physical appearance to most women is a man's level of authority or success.

At some unconscious level, women find men attractive who may be able to offer certain rewards and benefits that most other men cannot give. Above all else, women seem to prefer men who have resources. Resources could include money, the potential for the acquisition of money, advanced education, status among peers or even the ability to supply the basic life needs (home, food, transportation, clothing etc.) Resources, in short, are those things that allow you to have security in all of its many facets.

Women thrive and enjoy life when they experience security and stability. Women want security for themselves and for their children. If a man has resources a woman feels more comfortable in entering into a relationship with that man. Generally speaking, long-term relationships, like marriage, are best for women when compared to having many short-term relationships.

Women are attracted to men who are at or near the top of the pecking order in a group or organization.

An Educated Man

Women will look for potential if they cannot see immediate resource value in men. In other words, if a man currently has next to nothing to offer her, she will still consider the man attractive *if he has potential.* Does he have an advanced education? Does he have an area of excellence? Intelligence is still positively correlated with income and women know this on many levels. Bill Gates is not your likely vision of a dominant male, but he has more personal wealth and resources than some countries. There is a quip I heard someone say, "You know what a geek in high school is, right?" "No, what?" "A millionaire in 10 years." Welcome to the world in the 21st century. Brains have eclipsed brawn as the impetus for irresistible attraction, even though brawn is still interesting to women, and always will be.

An intelligent man is worthy of consideration even if he doesn't have a Ph.D. Intelligence is directly linked to being successful, shrewd, cunning, manipulative, and other traits that predict whether a man will develop resources.

There is another benefit to being with an intelligent man. Men who are not only street smart but are adept in the arts and sciences may have the added luxury of being interesting to be with!

> *Research indicates that women, on average, want to be with a man who is in the top 30% of intelligence among all men!*

A Tall Man

Women feel most comfortable when a man is about six feet tall. Women prefer tall men to short men. Most research indicates men from 5'11" to 6'2" will be held in the highest esteem among the majority of women. It's possible that over the course of human evolution, taller men held an advantage in subduing attacks on a home and therefore women seem to unconsciously gravitate toward the taller man. It's not easy to know precisely why women like taller men, but they do.

In the real world, where life isn't fair, tall men, on average, make more money than their shorter counterparts. Money in the 21st century is the currency of security and stability. Maybe at the end of the century, we will be using e-dollars instead of paper money, but the concept of value is not going to change. Women will always be attracted to men who can make their lives safe and secure.

A Man Who is a Little Older

Very few women marry a man who is younger than she is. Women seem to find men who are slightly older than they are attractive. In the United States, women tend to marry men who are about 4 years older than they are. A man who is a little older probably makes more money than someone who is their own age. An older man also probably has more stability in a career and has probably achieved more success than his younger counterpart. However, men who are *much* older are not perceived as attractive.

Significantly older men probably are not as attractive to most women for several reasons. First, significantly older men have less in common with the younger woman. A man who is 20 years older than a woman may have very different interests, hobbies and attitudes than the substantially younger woman. Second, a man who is much older is simply more likely to die than a somewhat younger man. This is important for a woman to consider because women (on average) prefer to be with one person rather than participate in a series of relationships with man after man.

Women, on average, find men who are about 4 years older than they are attractive.

A Man Who Treats Children Well

An interesting comment I often hear from women is that they like to see a man who is "good with kids." Every man who has children knows that women look at him more often when he has a child with him than when he is by himself. There is something magical that happens to a man's value in a woman's eyes when he is with a child. Women have an inborn need to see their children (present or future...or even past!) taken care of. Watching a man play or simply be with a child provides a woman with a certain inner knowledge that "this is a good man."

> *Women are attracted to men who are good with children, in part because this behavior is a predictor of commitment, kindness, and love.*

Kindness: The Power of Nice

If a man has a strong sense of personal, social and business mastery, then kindness is perceived as a an almost irresistible trait. A man who is "kind", but hasn't mastered the three spheres of living, will not be perceived as attractive by most women. Kindness is the trait that women look for when men are in public. How does this man act toward people he is with? What kindnesses does he exhibit now that will indicate how he will behave in 5 years? Women tend to think in the long-term when it comes to relationships, and this is one trait that women respond incredibly well to.

Women don't see a lot of "nice" in the real world and when they do, it means something to them. Women are extremely impressed by nice men. Displaying the signs of kindness can work to any man's advantage.

Men Who Commit

Women want commitment. They want to know that a man is interested in their future. Men can show a woman commitment in many different ways. Early in a relationship, commitment is seen by buying dinner or gifts. Most men can't afford to take a dozen women to dinner and shower a dozen women with gifts. If a woman feels like she is being taken care of, then she is likely to be interested in a man in the long-term. Commitment is often woven tightly with the emotion of love, as you might well expect.

It's hard for women to identify signs that any given man might be one who is willing to make a commitment. If they could gain access to information like this, it would help her cause. Women want a man who will commit.

What Happens If You Aren't Rich, On Top, Tall and Educated?

Is there any hope for the man who doesn't have a lot of resources? What if he doesn't have an advanced degree? There is plenty of hope, but you'll have to read the next chapter to see what happens next. Before we go there, let's look at women and see what it is that men really find enticing in a woman.

What Do Men Find Irresistible in Women?

What is it that draws a man to a woman? There is little that a woman possesses that can entice a man as much as her physical attractiveness. This doesn't mean that if a woman is not physically attractive that she has no opportunity to find happiness. It does mean that physical attractiveness is far and away the number one trait that men look for in women. No other individual trait comes close to a woman's appearance to make her irresistible to a man.

Men are driven, in large part, by their desire to have sex. This surprises no one. Napoleon Hill, history's greatest success philosopher, spelled out years ago that the most successful men in history have had enormous sex drives. However, men who pursue *only* their sex drives rarely become successful. A significant element in the success of any man is being able to channel that sexual energy into their work.

All of the care and decision-making that a woman must do before committing her energy into a man is something unparalleled by men. Men do not have the same long list of criteria that women do. Men generally do not find women attractive because of their wealth, their success, and their potential for achievement. These things are not powerful as an initial attraction to most men. Men are not typically that impressed or sexually attracted to women who have succeeded in business or done anything remarkable. The makeup of men is much different in attraction.

How Do We Know What Physical Attractiveness Means to Men?

In mathematical terms, men have evolved to seek out a woman whose waist to hip ratio is about .7. In other words, divide the waist measurement by the hip measurement and if the resulting number is between .6 and .8, a man will almost certainly find that woman instantly attractive. This is the bottom half of the hourglass figure. A man doesn't need a measuring tape to know what he is attracted to. He simply has a certain shape that he is designed to look for, and this is part of that shape. (Alternatively, women do have a filter for a man's physical appearance. Women tend to unconsciously filter out men whose waist is larger than their hips. However, physical attractiveness is not critical to a woman being attracted to a man.)

What is the Ideal Figure?

Some research indicates that the ideal measurements for a woman in a man's eyes are 5'5"', 120 pounds, 35-23-35. The measurements of preferred waist and hips seems to be consistent in the last 40 years. Ideal bust size seems to fluctuate slightly from decade to decade.

> *In addition to the hourglass body shape, a man is attracted to a woman who has a clear complexion, full lips, a smaller nose, larger eyes, arched eyebrows and a narrower jaw.*

Breasts have been the subject of fascination and even worship for millennia. How do people perceive various breast sizes? There is no question that larger breasts on a woman with a small or medium frame can draw attention. A man's perceptions of a woman with larger breasts is interesting to consider, though. In the United States, women with 34" busts have been rated highly on competence, ambition, modesty and morality. In the United States, big-busted women are perceived as less competent, less moral and less modest. Around the world however, big-busted women are highly valued, and men hold a negative view of women with small busts. In terms of attractiveness however, men are attracted to women with larger busts regardless of the negative perceptions that may be coupled with this attractiveness.

Generally, men are most attracted to young women. In Japan, men of all ages rate women at the age of 22 as the most attractive in their culture. In the United States that average age of a Playboy centerfold (usually chosen for sexual appeal, and not conversational style) is 21. Men prefer to marry women who are at least 3-4 years younger than they are. As men age and later re-marry, they tend to marry women who are significantly younger than they are.

TWO

Attraction for the Rest of Us

The measurements on the tape may not read 35-23-35 and the bank account may be only modest. You may not be 6' tall and you may not have the perfect education. Does that make us unattractive? Absolutely not! Not only is there hope, there are 13 proven strategies and behaviors that increase your attractiveness to others and only one of these 13 has to do with your physical appearance.

Focus On Others

What really gives you the charismatic edge over others is your ability to focus your attention toward other people. Learn to become fascinated by other people, their life experience and their work. This is a trait that people find simply irresistible.

People's eyes will gloss over when they hear what we call "I" language. When someone drones on about their successes and failures, their opinions and attitudes without gaining a similar amount of disclosure from others, they rapidly lose their "face value." When people talk without consideration for the thoughts of their communication partner, we say they are using, "I" language. They are simply talking about themselves and nothing else. Successful communication happens when people share equally about what is interesting and exciting to them. It entails learning more about what other people think and seeking additional information about other people's points of view.

The idea of bringing your attention to that of others is something that takes "getting used to." It's not always easy to really listen, understand and expand on what another person is talking to you about. The best habit you can develop is the avoidance of "me too" phrases. In other words when someone says,

"I just got back from Mexico."

You don't say, "I've been to Mexico too!"

(They don't care...yet.)

Instead say, "Really? Where did you go?"

"Ixtapa."

(You've been there, by the way...)

"Wow, what did you like about Ixtapa?"

"Oh, the beaches are unbelievable..."

Now, you can share your excitement. "I was there a few years ago and thought it was incredible. (Now direct the conversation back to the other person!) What did you get to do down there?"

You could have told all about *your* trip, *your* adventures, told them how beautiful the beaches were, etc. But that's what every boring person on earth does. People with charisma re-direct the conversation back to the other person so they can help the other person feel the fullness of the experience they are sharing. This is what focusing on others is all about.

Here is another example of keeping the focus on the other person when the temptation is to compare their experience with yours.

"I had a terrible day."

(You could say, "Me too, I got into a car accident.")

"Really, what happened?"

"My boss yelled at me and made me feel two inches high."

(You could say, "No big deal, ignore him." Or "Who cares what he thinks?")

"What an idiot. What got into him?"

"I don't know, he is normally a nice guy, but he was so mean today."

(Pursue the experience one more time.)

"So, what did he say?"

"He said that I was mis-treating customers. What an idiot."

"You wouldn't do that would you?"

(And on it goes...)

The temptation to compare this experience with one of yours is enormous. By avoiding the comparison and keeping your attention on the other person, they begin to find you charismatic because they get to feel the fullness of their emotions.

Talk about Their Interests

People become bored quickly if they are engaged in communication about something which they know nothing about. On the other hand, people find it stimulating when they find a kindred spirit, or at least an eager listener.

A useful overgeneralization is that men are interested in sports and women are not. Most women only follow sports to the degree they can't wrest the remote control from the hands of the male she is near. Therefore, it makes little sense to engage the majority of women in World Series and Super Bowl history. Similarly, the vast majority of men wouldn't recognize an astrological chart or a Meg Ryan movie it hit them in the face.

What does get people's energy up is when someone listens to them talk about what fascinates them most. Think of your favorite hobby, your favorite movie or TV show. Don't you just get "juiced" when someone will listen to you get excited about this? The people who really listen and get curious about our interests are those that we are attracted to the most.

It takes a sincere desire to become interested in some things. Your obvious interest in learning more about something *they* care deeply about is an instant rapport-building experience that could link two people together forever!

Keep These Off The Table!

Some subjects simply need to be left off the table early on in any kind of relationship. Other subjects are OK to broach but certainly could be more gently broached.

Men, of course, have absolutely no interest in a woman's past or current husbands, male friends and dating habits. It is simply not in the male psyche to listen to women discuss other men in any positive fashion. Men would rather discuss movies starring Diane Keaton than hear about past and present "other male experiences."

Women are not that excited to hear about a man's previous relationships either, but there is some amount of interest here that men do not experience. Women have a certain element of curiosity about a man's previous relationships. Women "dig deeper" than men do when trying to evaluate them. Nevertheless, men would find it wise to leave these topics off the table unless prodded by a woman.

Men tend to make mistakes early in relationships by disclosing too much direct information about their "resources." Women want to know this kind of information, but they don't want it in print on a resume. Women want to dig a little to get information, or, at least be offered bits of a man's resume.

Instead of telling a woman what his net worth is, he can disclose, for example, that he travels regularly, and let the woman discover more. It makes the journey more interesting and avoids being so "in your face." Women like discovering secrets, and if a man can withhold just a bit more than he has a tendency to, he can create stronger desire in a woman. This happens as a woman discovers bit by bit of information. However, as any fantasy role playing game player will tell you, part of the fun is being stopped at apparent dead ends, only to be given the magic words or key later. In short, taking your time in self-disclosure can be immensely helpful.

Look Appropriate to the Setting

A sport coat and tie probably is a poor choice of attire at a basketball game unless you are the coach. Jeans are probably out of the question at a nice restaurant.

To always be dressed in one style of clothing makes little sense if you want to appear attractive to those around you. The key really lies in being appropriate to the setting. It's not that you want to be just like everyone else, but there is one fact that always rings true in attraction, and that is "like attracts like". Dressing to the norm of the community, the office, the environment doesn't mean that you are just like everyone else inside, it means that the first time people see you, they see a little bit of themselves. That can be a valuable asset!

I recently hosted an event for an organization where single people would meet and mix with each other. I was wearing a coat and tie. I assumed that everyone coming to the event would want to be "dressed to kill." I was shocked when approximately half of the men showed up with messy hair and looking like they just got out of bed! I said nothing of course, but I was wondering what the thought process was walking out of the house for some of these people. I then really understood that there is yet another area that personal coaching has not yet addressed: Be aware of your self and those around you.

Care about Each Person in the Room

The "stuck-up person" is the one with their nose in the air. They clearly couldn't care less about anyone but themselves and, sadly, this is how they will end up in life. The qualities of empathy, caring, concern and genuine interest are keys to attracting high quality people. Those people who are truly attractive to others have learned how to not only focus externally (on others), but to develop a fondness for other people, especially those in close proximity. When you are around other people, do you find yourself opening doors for others? Do you perform random acts of kindness for people daily?

The trait of kindness is something that is hard to compete with, when it is seen as genuine. People who are kind are often seen as the most charismatic. Think about Princess Diana and Mother Theresa. These were two people of very different physical appearance and age. These two people sincerely cared for other people, and their depth of love and expressiveness earned them a special place in our hearts in the 1990's. How do you develop kindness and empathy for others?

The best answer is to actually engage in the activities that others appreciate. Do the little things that can make the lives of other people easier. You will always be at the top of other people's lists when you do.

Personal Mastery

Self-confidence is a trait that all people admire and are attracted to. Self confidence coupled with competence yields an obvious sense of personal mastery. The person with personal mastery has a fascinating mix of humility and pride that allows one to be certain and self-assured.

Personal mastery in a man is something women are drawn to like a magnet. There are few other traits that create a sense of certainty in a woman about a man other than his personal mastery. The reason is simple. If a man has competence and self-confidence, he can probably do anything in life and give her most anything she will desire. Competent people still have the ability to ask questions and learn. Not knowing the answers to all of life's questions means they still have a humility about them that is apparent. However, there is more. The competent person has learned breadth and depth about the areas he or she is excellent in. The person with a sense of personal mastery is someone who has no need to make others "wrong."

People with personal mastery know exactly where they are in life and they know where they are going. They have no delusions about themselves or others. They know that it takes a unique combination of hard work and wisdom to achieve in life. People with personal mastery never have all the answers to life's questions, but they ask great questions. They have great flexibility in their behavior and in their communication with other people. They obviously have a high level of awareness of the world around them.

Making Other People Feel Good

People who make other people feel good are not only helping others, they are helping themselves. When we see someone perform random acts of kindness, we see more than just a kind person, we see the highest good in our culture. Making other people feel good is something that anyone can learn to do, and something that everyone should engage in.

Little things tend to make people feel good. Simple positive comments about something as simple as a pin or the choice of clothing can make a person's day. A few kind and believable words about a presentation at a meeting or how someone handled themselves in a difficult situation bring appreciation and escalates rapport. Everyone likes to hear something good about themselves. We all need positive strokes from others and we rarely get these messages we so desperately crave.

It doesn't take much to bring out the best in other people. Too much flattery can be insincere and ineffectual. A few well-phrased words can go a remarkable distance in creating powerful attraction.

Enthusiasm's Power

What could be the single most important characteristic in charismatic people is that of enthusiasm. Watching a person who is excited about what they are talking about can ignite passion in other people. The word enthusiasm stems from the Greek, "en-tae theos," or "god within us." There is a certain "spirit" or "aura" that seems to fill people that are truly excited and passionate about their lives.

Almost everyone is captured by controlled enthusiasm. Enthusiasm that is frenetic can be exciting for a moment, but can be too much for extended periods of time. Controlling your enthusiasm, however, makes you smile and almost bubble with vitality.

Enthusiasm can be experienced when we are with someone who brings out the best in us. It also is an internal response to having goals and desires that we are working toward fulfilling. People who know where they are going in life seem to captivate the atmosphere when they enter a room. You can be this person if you begin to think about where you are going in life and consider the steps you are going to take to get there. Do you have a clear vision of where you want to be in life? If so, you are instantly more attractive to everyone around you!

Be Healthy

People who are vibrant and attractive to others are those that have sound mental and physical health. If you don't, go see a medical professional to help you get there. For most people, it's fairly easy to improve your health. Taking simple steps like losing weight, quitting smoking and eating more fruits and vegetables improve your physical appearance. Doing these simple things also helps you have more energy and makes you feel better. When you feel better, you are more exciting to be with for others.

I've learned the hard way over the years that balance is important. You must work hard, but you must also get the rest your body needs to recharge its batteries. Some of the things I notice people could do to help themselves include drink less alcohol, avoid non-prescription drugs and see their medical doctor far more regularly.

Did you know that people who see their M.D. regularly live longer? It's true. The reason is simple, of course. You see your doctor and you keep your body in tune just like seeing your car mechanic for maintenance keeps your automobile in tune. If you are healthy, you find you have energy, and people are attracted to and by your energy.

Develop Certainty

Constantly self-educate. The more you know, the more you can share with other people and engage in their favorite experiences. Become a voracious reader. Travel if you can. Learn from audio and video resources about your areas of interest. As you become more knowledgeable about various subjects, people seek you out for ideas and advice because you are an expert. Experts also know how to be wrong, and how important admitting to being wrong can be. They also know that people search them out because of the confidence that their certainty brings them. People with certainty about who they are and where they are going are magnificently exciting to be with. We are inspired by their focus and their quest. Certainty is not arrogance. Certainty is an almost faith-like conviction that someone experiences about her path in life.

People who have this sense of certainty are easy to spot. They take an enormous amount of action toward their goals and objectives in life. They have faith in themselves and see almost every experience in life as moving them closer to where they are going.

People who learn more about the areas of their personal excellence are fun and interesting to be with. They are the do-ers in the world. They learn and they apply what they learn into their lives because they KNOW they will succeed.

Orientation to Action

Very little is more exciting to us than being with people who make things happen. Action-oriented people are extremely attractive. Everyone likes to be with someone who can make something out of an idea or a thought. Action-oriented people make lots of decisions, many of them wrong, but they are constantly moving and that creates energy.

Maybe what is most attractive about people who take action in all facets of life is the fact that they have overcome procrastination and general inertia. People who take action are perceived by others as successful (or potentially successful) and that is one perception that causes attraction!

Examples of people in motion are people who like to play various sports, people who do work they love and those who make decisions, and then follow through on them. The motion needs to be focused, however. A person who moves but gets nothing done is not interesting to be around. The action you take must be directed and focused.

Personality Plus

Personality does matter. We all tend to admire and be drawn to certain personality traits and driven away from others. There are many characteristics that blend to create a fantastic personality. Let's look at four of the cornerstone traits: **Tolerance, a Sense of Humor, Sincerity and Flexibility.**

People who have a tolerance of others tend to be more charismatic. Tolerance doesn't mean that if a person approves of criminal behavior that they are tolerant. That is stupidity. Tolerance means that differences are perceived as nothing more than differences between people.

The majority of people are also drawn to those who have a sense of humor. A person who can laugh at himself is someone who might be a lot of fun to be around. Anyone can make fun of someone else, but to be able to be self-effacing with style is something that is admired. People who can make light of somewhat difficult events can draw the attention and admiration of others.

Sincerity means a lot to people. Women almost equate sincerity with long-term commitment, which is what most women biologically need. Sincerity is not always easy to manifest in relationships because of the very different ways that men and women think. People of both sexes admire those people who can be upfront about who they are and what they feel when it comes to the long-term.

Flexibility is the trait of adaptability, which is, of course, what life is all about. The person who can adapt the best to changing circumstances is usually the most successful or effective individual in any given situation. People admire flexible people and find them enticing to be with.

The Talking Body

Some people just sit there when they talk. Their eyes don't light up, their hands never leave the table, and their voice never rises or falls. They are not attractive, not interesting and not going to be the focus of anyone's attention.

People who are exciting, intense and fun - move. The most charismatic people of the 20th century were people like Elvis Presley and Marilyn Monroe. When they moved, the whole world watched.

Does your face move when your mood changes? Does your body move as you become excited? If not, start paying attention to the people who are drawing the attention you desire and see what they are doing right!

Can You Hear Your Body Talking?

Your body language and your physical appearance will boost you to the top or keep you among the majority of people in the middle and bottom. You have less than 10 seconds and, realistically, closer to four seconds to make a good first impression on those you come into contact with. There is a world of research that clearly indicates that you will be judged professionally and personally in the first few seconds of your meeting someone for the first time. In fact, your first impression is recorded and is used as a yardstick for all future communication by those whom you meet. Whatever that first impression is going to be on your part, you want it to be intentional and on purpose.

Before going any further in discussing verbal communication, we better take a look at how to really be perceived as attractive with your body language. Most people are completely unaware of just how much their body says and how it often contradicts what their words are saying! There are numerous elements of what we call body language. They include your physical features, both changeable and unchangeable, your gestures and signals that you send to others at the conscious and unconscious level, and the space that you use when communicating with others. In this chapter, we will touch on all of these important areas of body language.

What You Look Like Really Talks

Let's begin with your physical appearance. Here are some astounding facts that will give you pause for thought when you consider how important appearance is in attraction.

Your perceived level of attractiveness by other people will be a significant benefit or detriment in your life. It isn't fair, but it is true. People who are attractive learn how to deal with less than perfect physical features and work with what they can to best advantage. Before we consider just how to increase your face value in the next chapter, look at the results of some fascinating studies about physical appearance.

The Power of Physical Appearance

Did you know that in university settings, professors who are considered physically attractive by students are considered to be better teachers on the whole than unattractive professors? Attractive professors are also more likely to be asked for help on problems. These same attractive professors also tend to receive positive recommendations from other students to take their classes and also are less likely to receive the blame when a student receives a failing grade! (Romano and Bordieri 1989)

Did you know that marriage and dating decisions are often made with great weight placed on physical attractiveness? A wide variety of research indicates that men will often reject women who are lacking (in their opinion) in positive physical features. Women, on the other hand, place less significance on a man's physical attractiveness in considering him for a date or marriage. (studies by R.E. Baber)

Did you know that in studies done on college campuses, it has been proven that attractive females (attraction as perceived by the professors) receive significantly higher grades than male students or relatively unattractive females? (studies by J.E. Singer)

There is even more evidence that shows you must make the most of what you have physically.

Among strangers, individuals perceived as unattractive in physical appearance are generally undesirable for any interpersonal relationship! (studies by D. Byrne, O. London, K. Reeves)

In one significant study of 58 unacquainted men and women in a social setting, we learned that after a first date, 89% of the people who wanted a second date decided to do so because of physical attractiveness of the partner! (Brislin and Lewis)

In the persuasion process, attractive females are far more convincing than females perceived as unattractive. (Mills and Aronson)

Among American women, the size of a woman's bust is significant to how both American men and women perceive the woman. Women with a "medium" sized bust are considered to be more likable and have greater personal appeal than women with a large or small bust. Women with a small bust are perceived as more intelligent, competent, and moral. Women with a large bust are generally perceived as less intelligent and less competent. (Kleinke, Staneski, 1980)

In yet another study, we find that young men who are obese are generally considered to be slothful and lazy. Both men and women who are obese are generally perceived to have personality characteristics that place them at a disadvantage in social and business settings. (Worsley, 1981)

Can You Change Your Appearance?

Study after study reveals that how you look is critical to someone's first impression of you. So what can you do to change how you look? You can't change everything about your physical appearance but you can definitely make changes that will give you a booster shot.

Research studies tell us that the "exposure principle" increases our "face value." Specifically, the exposure principle says that the more often you are seen by someone, the more attractive and intelligent you appear to them. If you weren't gifted with a Cindy Crawford or Tom Cruise face, then it's time for you to take advantage of the exposure principle.

If you don't have the advantage of being "seen" time after time by a person or a group, then you must make the most of what you have. In other words, you want to look as good as you possibly can on every given day. Because of the significance of body image and weight, you must do what you can to keep your body weight down and your body in shape for your overall image to be as good as it can be.

Your teeth will tell a tale as well. If your teeth are yellow and look like you just ate, your face value is obviously greatly reduced. Do everything you can to keep your teeth pearly white, and you will be perceived as more attractive. (You've already seen the benefits of the perception of attractiveness.) When you watch the news tonight on TV, look at the teeth of every news anchor, weather person and sports announcer. They all have beautiful white teeth. There's a reason for that, and that is positive impression management.

Where You Sit Can Change How People Look at You!

Standing awkwardly in someone's office can be a problem that needs an immediate solution. As soon as pleasantries are exchanged, you and your customer should be seated. If you are both standing for an extended period of time and your customer doesn't have the forethought to offer you a chair, then you can ask, "Should we sit down and be comfortable?" Unless you are in a retail environment, sales are not made and deals are not negotiated standing up.

You may have an option of considering where to sit. If so, you are in luck. Scientific research is on your side in telling you exactly where to sit. You will often have the choice of where to sit on lunch or dinner dates at a restaurant, or in business settings such as in a meeting room. If you are in a restaurant, quickly search out a location that allows you to sit facing the people in the restaurant so your client is obligated to sit facing you, away from the clientele and staff of the restaurant. Booth seating can be ideal to achieve this.

Your partner's or client's attention should be on you, not the waitress, bus boy and the dozens of other people in the restaurant. Your seat selection will assure you his attention. Once you have the attention of your customer, only then can you effectively begin your presentation or engage in conversation.

How Do You Select Seating?

Ideally, you can create a seating arrangement that is most likely to facilitate the communication process. Here are the key rules in seating selection.

1) As a rule, if you have already met your client or friend once and you know they are right-handed, attempt to sit to his right. If she is left-handed, sit to her left.

2) If you are a woman attempting to communicate effectively with another woman, sitting opposite of each other is as good or better than sitting at a right angle.

3) If you are a woman attempting to persuade a man to your way of thinking, the best option is to be at a right angle to him, if at all possible.

4) If you are a man attempting to persuade a man, you should be seated across from each other in the booth setting, if possible.

5) If you are a man attempting to communicate well with a female in business or in a social setting, you should be seated across from her at a smaller, more intimate table.

What Do You Do Once You are Seated?

Waiting for the waitress to come to the table in a restaurant can be awkward if you do not know your date or your client very well. If you are meeting your client in her office, you will immediately get down to business after brief pleasantries. (It should be noted that sometimes pleasantries do NOT have to be brief. Many of my biggest and best presentations were made in the last two minutes of meetings that would extend up to two hours in length discussing everything from baseball to sex to religion. The level of rapport and quality of mutual interests will ultimately be your guide.)

Once seated, keep your hands away from your face and hair. There is nothing good that your fingers can do above your neck while you are meeting with a client. The best salespeople in the world have wonderful and intentional control of their gestures. They know, for example, that when their hands are further from their body than their elbows that they are going to be perceived as more flamboyant.

While you are seated, if you are unfamiliar with your date or client, it is best that you keep both feet on the floor. This helps you maintain control and good body posture. People that are constantly crossing and un-crossing their feet and legs are perceived as less credible. And people who keep one foot on their other knee when talking have a tendency to shake the free foot, creating a silly looking distraction. Feet belong on the floor.

Meanwhile, your hands will say a great deal about your comfort level. If you are picking at the fingers of one hand with the other, you are tending to reveal fear or discomfort. This is picked up by the unconscious mind of the customer, and makes *her* feel uncomfortable. If you don't know what to do with your hands and you are female, cup your right hand face down into your left hand, which is face up. Don't squeeze your hands; simply let them lay together on your lap.

For men, the best thing to do is to keep your hands separate unless you begin to fidget, at which point, you will follow the advice of your female counterpart, noted above.

How Close is *TOO* Close?

Whether seated or standing, you should stay out of your client's "intimate space." Intimate space is normally defined as an 18-inch bubble around the entire body of your client. Entering this space is done so at your own risk. This doesn't mean that you can't share a secret with your date or your client. This doesn't mean you can't touch your date or your client. It does mean that if you enter into "intimate space," you are doing so strategically and with a specific intention. There can be great rewards when entering intimate space, but there are also great risks, so be thoughtful about your client's "space."

Similarly, if you leave the "casual-personal" space of a client, which is 19 inches to 4 feet, you also stand at risk of losing the focus of attention from your client or counterpart. Ideally, most of your communication with a new customer should be at the two- to four-feet distance, measuring nose to nose. This is appropriate, and generally you begin communication at the 4' perimeter of space, and slowly move closer as you build rapport with your client.

What is *Effective* Eye Contact?

Eye contact is critical in any face-to-face meeting. As a rule of thumb, you should maintain eye contact with your client 2/3 of the time and with a date about 80% of the time. This doesn't mean that you look at her eyes for 20 minutes then away for 10 minutes. It does mean that you keep in touch for about seven seconds, then away for about three seconds, or in touch for about 14 seconds and away for about six seconds. Eye contact doesn't mean just gazing into the eyes. Eye contact is considered any contact in the "eye-nose" triangle. If you create a triangle from the two eyes to the nose of the customer, you create the "eye-nose" triangle. This is the area that you want 65-70% of eye contact.

Should you sense that your client is uncomfortable at this level, reduce your eye contact. Many Americans who were born and raised in the eastern countries (Japan, for example) are not accustomed to the eye contact that Americans are.

Eyes are a fascinating part of the human body. When a person finds someone or something very appealing, their pupil size (the black part of the eyes) grows significantly larger. This is one of the few parts of body language that is absolutely uncontrollable by the conscious mind. You simply cannot control your pupil size. If you are interested in someone else, your pupil size will grow dramatically. If someone else is interested in you, their pupils will grow larger when looking at you, and there is nothing they can do about it. This is one of the powerful predictors of liking in nonverbal communication.

It should be noted that pupil size will also get larger in situations of extreme fear and when a setting is dark. Pupils expand to let more light in and, like a camera, when the setting is very well lit, the pupils will contract to the size of a very tiny dot. You'll read about this in more detail in the next section.

If you follow the tips in this chapter for improving your appearance, being careful about appropriate dress and are careful with your use of space, you will be perceived as more attractive in personal relationships and in business.

There are two other telling behaviors relating to the eyes.

First, if someone is blinking far more rapidly than they normally do, that is usually an indicator of one of the following three things. The most benign is that they find the lighting in the space annoying. But, it can also be an indicator that the person with whom you are communicating is experiencing a high level of anxiety. Or, it can even mean they are lying. In 1998, President Clinton gave a short speech offering his reasons for having an illicit affair with Monica Lewinsky. During this speech, his eyes blinked a momentous 120 times per minute. Two days later, he gave a speech about a U.S. bombing raid on a terrorist group overseas. In this speech his eye blinks were about 35 per minute. The difference is extremely important in evaluating the comfort level and honesty of the President in each situation. If someone is blinking far more often than normal (and you do have to know what normal is for each person you meet and adjust for lighting), you know they are very probably extremely anxious and very possibly lying.

EYE CONTACT

Second, if you are in conversation with someone and their eyes are easily distracted by the goings on in the environment, this is usually a good indicator that you haven't earned the interest of your listener. In general it is a very wise strategy for you to keep your eyes well trained on your date or business associate in distracting environments. To constantly look around at the environment when you are with someone else is perceived as rude. To keep eye contact with another person instead of being distracted by extraneous activity is considered flattering and complimentary, especially by women.

So, there you have it. You don't have to look perfect and own Trump Towers to be incredibly attractive to the multitudes! However, you want to take advantage of every aspect of your attractiveness that you can. And later in this book, you will discover specifically what to do to really bring your best you forward.

The Eyes Have It

Did you know that you are able to get a pretty good idea of how someone feels about you by looking at his or her eyes? You get even more information about how someone feels about you when you put that "look" into the context of their facial expression and their body language.

Women initiate about 65% of all flirtatious encounters with men. Usually this is done with their eyes. How you look at someone can be perceived as seductive, frightening, caring, loving, bored, secretive or even condescending. The eyes reveal a great deal about what is going on inside of us. If you can learn how to look and send the right message at the right time with your eyes, you will be perceived as more attractive by more people.

There are six basic emotions in the human experience and the eyes capture them all. There are many more than six different emotions, but most of the emotions we experience are a combination of the six basic emotions. By simply looking at a person's eyes, we can tell whether they are experiencing, **happiness, surprise, disgust, fear, anger or sadness**. Think about that. Across the world, people are the same in this respect. We all show the six basic emotions in the same fashion. The eyes are amazing windows to the emotions we all experience. By paying close attention to the eyes, we can learn a great deal about people and, in particular, those we wish to attract.

It is a true statement that most people will judge other people in the first two or three seconds after their first meeting. Therefore, doesn't it make sense to have them hypnotized by your eyes and your understanding of their wants and needs? How do you do this? You use your eyes in simple yet powerful ways to build rapport and create feelings of arousal in the person you are attempting to attract. To do this, you need only to apply the key ideas you will learn in this chapter.

I (KH) recently had laser surgery on my eyes to improve my vision without glasses. In the screening process, I learned that some people shouldn't have the surgery because their pupils dilate (get bigger and blacker!) to a size that is abnormally large. Everyone's pupils dilate when it is darker in the environment and they contract when it is lighter. When the sun is shining brightly in your eyes, your pupils will be at their smallest. When you walk into a dark room, your pupils will be at their largest. The pupils get larger to gather more light. This helps the eyes see more of what is in the environment.

Your pupils will also get larger when you are terrified. There is an evolutionary response in your body that helps you collect more information about an experience that is frightening. The senses all sharpen in moments of great fear. Your hearing becomes more acute, your sense of touch is enhanced, and you can even taste fear. The pupils in your eyes get larger. This helps bring more light in, even if the environment is already well lit. Your brain needs that information to help you escape and to protect you from danger.

Everyone's pupils dilate to a different maximum size, and everyone's pupils have a slightly different normal state. However, there is one amazing fact about those eyes: When someone looks at you and their pupils get big and black, they are either scared to death of you, or they like you!

It's almost impossible to control the increase in pupil size that occurs when we see something we like. This expansion is also an evolutionary process that happens to enable you to take in more of something that is very dear you. For the observing person, knowing this is an uncontrollable response can make it difficult to determine the cause.

Recent research into pupil dilation has proven quite interesting to the field of attraction. When a researcher showed pictures of a baby to women, the results showed most women experienced measurable pupil dilation. When the pictures contained a mother and baby, this elicited an even greater pupil dilation response. These same women viewing a beautiful landscape experience an enlarging pupil size as well. Interestingly, women viewing a picture of an attractive man, on average, don't experience quite the size of pupil dilation noted in the above scenarios! Women can be impressed by a man's appearance, but at least at an evolutionary or biological level, physical appearance isn't going to turn on every woman who passes. (Just what does turn women "on" will be discussed later in this book.)

These same researchers took the picture of that same beautiful baby and showed it to men. The men's response was a non-event. Their pupils, on average, didn't dilate. When viewing the baby with the mother there was again, a non-event. Generally speaking, nothing happened. When the men were shown pictures of a beautiful landscape, again, nothing happened. As soon as a man was shown a photo of a beautiful woman, however, the pupils, on average, dilated to a big, black orb. A man, it would appear, is very much turned "on" by the sight of the beautiful woman, even just a picture of one.

Pupil dilation experienced by women, when in the presence of real-live men, is another matter. Women typically are not visually aroused by photographs in the same way that men are. Women are very stimulated by some men in some contexts. When women are sitting across from men who arouse them, their pupils do dilate. To the observant witness, it is obvious. Most people are oblivious to the enhanced pupil size, and *yet it is one of the most telling signals of attraction.*

As a public speaker, I (KH) have talked to thousands of audiences all over the world. As I speak, I am aware of the women whose eyes are big and black, and I always address my presentation to them, making eye contact with those who appear to be aroused or attracted to me. They don't know this is why I selected them to make eye contact with (at least they didn't until now). Part of my job is to excite and inspire an audience when I speak. Therefore, I need to gain as much rapport with the audience as I can. By making contact with the people who like me the most, I am able to gain agreement from those people. They nod their heads, lean forward, show interest, smile, and everyone in the audience sees how much fun they are having. In groups, head nods are like a virus. Once one person nods his head, almost everyone does!

I receive all of this positive feedback, in part, because I don't just look at faces in an audience. It is because I look at the eyes of dozens of people in the audience and find the biggest pupils I can locate! These searches are like a treasure hunt that has a pot of gold at the end. If I can do this with an audience of 50 or 100, can you imagine how easy this is to do in a smaller group at a party or in a public place? Start paying attention to the eyes that are looking at you.

You may wonder, "What if you are wrong? What if those eyes are just big because they are among the women whose eyes are normally large? Then aren't you just fooling yourself into believing that all of those women are attracted to you?" My response is, "Of course." When you hallucinate, it should always be something that increases your self-esteem and self-confidence! We'll talk about how your beliefs and self-confidence affect your attractiveness elsewhere in this book!

Recently there was a fascinating study which revealed that when you show two pictures of the same woman to a man, the man will perceive the picture of the woman with the bigger pupils to be significantly more attractive. Many magazine cover editors know this and actually touch up the cover picture to show larger pupils. This makes the final picture irresistible to the magazine purchaser. We simply love people with big eyes!

Eye Contact

Men often ask, "How can you tell when a woman is interested in you?" My response is simple. "If you see her give you one glance, she saw you. If she looks back in less than a minute, she finds you attractive."

A man can stare at a women for 30 minutes or more and never have their gaze reciprocated. Meanwhile, every woman in the environment can see who the man is interested in and direct their attention elsewhere! Women, on the other hand, are more intuitive about eye contact. They will look around a room and see who is there. They will give second glances to those they are attracted to and avoid men they are not interested in.

When eye contact is made, it is a good idea to give someone "an eyebrow flash." This is a quick "raising" of your eyebrows that lets the other person know that you are attracted to and interested in them. **The eyebrow flash is common to every culture on earth**. People who do not reciprocate with an eyebrow flash are

60

sending a message that they are not interested. Make sure if someone "flashes" you, that you flash back!

After the flash has happened, you will find it uncomfortable to maintain eye contact with that person for more than a few seconds. Therefore, break your eye contact after a few seconds, look down and then back again at the person you are interested in. Look down, not up. Looking up is usually a sign of not being interested in the other person, so be careful not to do that!

Eventually, you strike up a conversation with someone. You find yourself in one-on-one communication with another person with whom you are initially attracted to. How can you use your eyes to enhance the likelihood they will be even more attracted to you?

The amount of eye contact and the type of eye contact we have with another person is important to attraction. Women have a variety of responses to lengthy eye contact. Most women love to be the only person a man will look at in a room. They want undivided attention and are aroused when they receive it. However, there are a percentage of women who have a fear of being dominated or being harmed by men. The roots of these fears usually stem from a time when they were younger. They may have been harmed or abused by a man. These women do not feel comfortable with lengthy eye contact. Therefore what turns "on" one woman can quickly turn another "off".

Men, on the other hand, tend to find exclusive eye contact very arousing. Men rarely have fears related to eye contact when they are with a woman they are attracted to. Men and women both want to be the center of the other person's world. In fact, the most charismatic and charming people are those who can make the world melt away around another person.

You know how it feels when you are with another person in a busy location, and you felt like you were the only person that your partner was even remotely interested in. In being irresistibly attractive, you will always want your companions to feel like that. How can you do this, though, if you sense the other person doesn't want to feel "stared at?"

Use the 70% rule in the United States. 70% of the time, you will look at the other person in the "eye triangle." As discussed earlier, this triangle extends from the ends of the eyebrows to the tip of the person's nose. Caress your partner with your eyes as you gaze in this triangle. When you break eye contact, do not break to look at another person. Keep your focus of attention with this person. When you intentionally break your eye contact, do so by looking down, to the left, or to the right. Looking up in response to a question or while telling a story is just fine, but looking up to break eye contact is often thought of as a sign of waning interest!

Another way to break eye contact is to move your eyes outside of the triangle and move your eyes to the person's hair, compliment the person (only women in this case) on how nice their hair looks, and then return to their eyes. On a first date, a man should use what I call the "shoulders rule." A man should only gaze at everything that appears above the shoulders on a first date. Men find it almost impossible to avoid looking at the more curvaceous parts of a woman's body. That's the biology that we have been given even after millions of years of evolution. Most women, however, would appreciate being seen as more than just sexually attractive on that first meeting. They want to feel sexy, they want to look sexy, they want you to notice; yet they don't want you to get carried away with it! It's an interesting paradox. Women spend 2-10 times more time getting ready to go into public where they feel they may find an attractive mate. Women will look as good as they can and they know that their physical appearance is what will draw men to them. However, they also want much more than to be just eye-candy for a man.

Men desperately want eye contact with women. Men gauge the interest of a woman by her eye contact. Men are very competitive and territorial when it comes to women looking at other men. They see this as a sign that the woman is no longer interested in them, or that the interest is fading. Therefore, if a woman wants to continue to attract the man, the woman needs to maintain steadfast eye contact. A man's self esteem will crumble if a woman begins to observe all the male competition when in the presence of a man.

On the other hand, we can safely predict that if we have the full attention of the one we are with, they hold us in esteem to some degree. There is no other indicator that is as powerful as eye contact that can show interest in another person. Our eyes unconsciously and automatically move toward that which interests or arouses us. We all know that and we all judge our value in some part by the response we receive from other people. The eye contact we give and receive is just the beginning of attraction.

It's interesting to note that people with blue eyes are more demanding of eye contact than people with brown eyes. It is quite easy for us to look at a person with blue eyes and see the size of their pupils. When they expand and contract, it is evident. The person with blue eyes is used to people looking at them for an extended amount of time, in part, because of the contrast between their blue eyes and black pupils. The contrast can be striking at an unconscious level.

People with brown eyes, on the other hand, are used to other people looking away more rapidly because at the unconscious level it appears that the person with brown eyes is not as interested in them. The brown eyes present a weaker contrast to the black pupils. It often appears at the unconscious level that those brown eyes are not interested in us! Therefore we tend to look away from the person with brown eyes, when in fact, they may have been very interested in us.

When the person you are attracted to has brown eyes, you must pay more attention to their eyes to see the contrast between the black and the brown. What seemed to be an uninterested person may be someone who is actually quite excited about you!

Confirming our beliefs about the value of eye contact in attraction is a study that was done some years ago. People watched films of a couple that communicated with each other in two distinct ways. The first film showed a couple that had eye contact during 80% of their communication. The second film showed a couple that had eye contact 15% of the time. The observers of the films rated the couples that had eye contact 15% of the time as *cold, cautious, submissive, evasive, defensive and immature* (among others). The observers of the films whose couples had eye contact 80% of the time described the people in the film as *mature, friendly, self confident, sincere and natural.*

Gazing into someone's eyes is much more than just something special for the two engaging in the eye behavior. It is a clear signal to the rest of the world. These people like each other!

The Eyes Don't Lie

Whenever you are in a situation where attraction takes place, there is plenty of room for deception!

People have been known to stretch the truth about their age and income, their intentions and even their degree of love for another. The eyes act as a leading indicator of truth and deception.

In 1997 and 1998 I was invited to participate on hundreds of radio shows talking about the body language of President Clinton, Monica Lewinsky, Kathleen Wiley, Hillary Clinton and numerous other key players in the White House scandal that led to the President's impeachment. The interviewers wanted to know who was telling the truth, who was lying and what the facts were based on the body language cues I was reading.

Having carefully watched the President for almost 7 years, I was familiar with his every facial expression and body posture. President Clinton certainly was the most charismatic president since John Kennedy. His ability to excite an audience and win over people who disagreed with him is legendary. He is an outstanding speaker who thrives on being in the limelight. There were, however, two speeches and the famous grand jury testimony where the President was not his usual charismatic self. On these three occasions, he was uncomfortable about the deception he felt it necessary to accomplish. The first was when he shook his right finger at the world and said, "I did not have sex with that woman, Monica Lewinsky." The next was during the grand jury testimony where he was videotaped from the White House. The third was the speech he gave that very evening, after the grand jury testimony, when he offered his regret for being involved in the situation. On these three occasions, his eyes gave him away as being deceptive. The one speech that I want to share information with you about is the speech where he apologized for his behavior.

Throughout Clinton's Presidency, I watched him communicate with the country, and even though has been called "Slick Willie," his body language has rarely indicated any internal discomfort with what he communicated to the public. In this particular "apology speech" however, his anxiety, fear and deception cues were very high.

When I watch someone to see if they are being deceptive, I look to the eyes for important cues. I want to especially know how many "eye blinks" per minute a person experiences in contrast to when they are telling the truth. For 7 years President Clinton's "eye blink" pattern is that of about 7-12 blinks per minute. That is very normal. During the "apology speech" however, his eye blinking was recorded at 70 per minute! What that means is that on some level, the President was being deceptive in his communication.

Once eye irritants like contact lenses and allergies are ruled out, the only internal experience that will cause eyes to blink at that pace is the experience of anxiety normally associated with deception.

You should know that some people have eyes that never blink and a small number of people have eye tics that just won't stop blinking. On average though, a person will blink from 7-15 times per minute. When a person is being deceptive, their eyes will blink 5-12 times that pace. Like pupil dilation, controlling eye blinks is very difficult if not impossible. Take a moment right here and now. Simply try and keep your eyes open for 30 seconds without blinking. It's not easy is it? Now here's another experiment for you to do. Stare at a friend for 30 seconds. No blinking is allowed.

It is very difficult to stop your eyes from blinking! If you are in conversation where someone is telling you about something, and suddenly you notice a big jump in the number of eye blinks per minute, you can safely bet there is some deceptive behavior going on somewhere in what they are saying!

The eyes may or may not be the windows to the soul, but they certainly are strongly linked to the emotions and the entire make up of the brain's responses to other people.

Sound Bytes from Scientific Research

- Generally speaking, the longer the eye contact between two people, the greater the intimacy that is felt inside.
- Attraction increases as mutual gazing increases.
- Others rarely interrupt two people engaged in a conversation if they have consistent eye contact.
- Pupils also enlarge when people are talking about things that bring them joy or happiness. They often contract when discussing issues that bring them sadness.
- Women are better non-verbal communicators than men. Men can improve, though. One reason men aren't as good in reading body language is that men often communicate sitting or standing side-by-side and don't see as much non-verbal communication as women do.
- Women engage in more eye contact than men do.
- Eye contact has been shown to be a significant factor in the persuasion process.
- When women are engaged in a high degree of eye contact, they tend to be more self-disclosing about personal subjects.
- When eye contact decreases, men tend to disclose more and women tend to disclose less!
- The longer your eye contact, the more self-esteem you are perceived to have.

- The more eye contact you can maintain, the higher self-esteem you actually rate your self on!

Simply Irresistible Eyes

Given what we know about the eyes and attraction, we can summarize the experiences of millions of people into a few key ideas for irresistible attraction.

- Start with your eyes. Are they clear, or are they bloodshot? People who look at you will notice and the clearer your eyes, the more attractive people will perceive you to be.
- If you wear sunglasses, get ready to take them off. They can add mystery and they can ultimately be a big turn off. People want to see what they are getting. They want to see your eyes.
- If you wear glasses, consider contacts or other alternatives. People need to be able to see your eyes!
- If you want to attract someone, look at him or her. Look at them again and again. And smile!
- Look at a man from head to toe on the initial contact. He will be flattered. Look at the woman from the shoulders up, and she will think you have depth and possibilities.
- Look at the person you are attracted to about 70% of the time when you are communicating with them.
- Avoid looking at others for any length of time when you are with someone who may be special. Make the person feel as if they are the only one in the room that could possibly catch your eye.

- Remember that the longer they have eye contact with you, the more emotional arousal they are experiencing inside!

THREE

Initial Impressions

First glance. That's it. That's about all you have to make your first impression on another person. They will either be interested or not. You can change that impression later, if you get a further opportunity to do that. However, there will always be something in the back of their mind that remembers that first moment they saw you.

That might be a good thing. They might have caught you doing a heroic act and will always remember that incident no matter what you do after that. Or, they may have caught you at your worst, and will have to overcome that impression, always trying to rectify that against what they later learn about you that is positive.

> *You have two to three seconds to impress a person. They will either be attracted, or they won't.*

If you still are not convinced, go to the mall or to a bar or a restaurant. Look around at everyone there. How long do your eyes rest on that other person before you have made a judgment about them?

Are they attractive? Are they interesting? Are they worth the effort to possibly get to know them?

It doesn't take long to make that assessment, does it?

When you look around the room, what is it that really strikes you about a person? What is it that you notice first?

When scanning that room, how many of them do you not even see? That may sound odd, but it's true. We have filters in our minds that actually filter people, objects and events out of our minds.

It is like looking through a perforated shield. Some of "reality" is completely blocked out of vision. This is a convenience to the conscious mind as it narrows the decision-making process and relieves us of the effort. In our busy world, there is just too much information coming at us in any given moment. If we had to try to consciously determine which of those events or objects to pay attention to in each and every moment, we would be exhausted and have no energy or brain space left over to do the things that we deem important.

Have you ever been out with a friend and they make reference to someone, only to find out that you didn't even realize that person was there? Before we even get to the phase of sorting who is interesting and who is not, an unconscious filtering system has already narrowed the field. People that are completely indistinct to us, or that offer no interesting possibilities, may be completely shut out of our field of vision. For teens, that may be middle-aged to elderly people, for instance. They may find "no use" for them, and therefore don't even see them consciously.

So our unconscious mind conveniently filters out the things that it has determined to be unnecessary at the moment. Of course, it could be wrong!

What is left for us to do is to look through the remaining holes in that perforated screen. We allow ourselves to pay attention to the narrowed field of vision. In doing so, we then begin the sorting process that tells us who is interesting and attractive, and who is not.

Each of us has these filtering and sorting systems that we use to hone in to possible mates, potential friends, and people we want to be associated with. We will have different criteria for each of these categories.

If we are seeking out business associates, we will have a certain system for detecting who will offer a promising connection. We may be attracted to someone who appears to be self-confident, in control, and professional.

When we are making new friends, our sorting factors are different. We may be attracted to people who appear similar to ourselves. Those who exhibit similar tastes in clothes, hairstyle, hobbies and interests, marital status, and economic level would more likely come into view.

Yet, when we are seeking out a romantic relationship, we have an entirely unique system of sorting. It is based on our past experiences, present conditions, and hopes and expectations for our future. Generally, when we are searching for the right mate or life partner, our guidelines become more specific and detailed.

Regardless of our specific mating criteria, the impression we get of others will be instantaneous. We filter out anyone who doesn't meet our standards.

A man who has been "cleaned out" in a divorce, a woman who is coming out of an abusive relationship, or a teenager who is looking for their first love are all going to have individualized filtering systems. And, as we move through life, our own criteria for sorting will change. As we mature, as we gain more experience, whether positive or negative, we adjust our assessment of what is beneficial and desirable to us, and what needs to be avoided.

Beyond what was earlier described as our genetic wiring for attraction and desire, there are fundamental criteria that will dictate the reaction we have when we first encounter a stranger. Our genetic wiring says that men will seek out women based on the availability of a sexual encounter, and that women will seek out a man based on potential economic security. That may be how we are hardwired and be representative of our basic animal instincts. However, in the modern world, in social settings, through time and aging, and with all human ecological factors considered, there is still the "first glance".

The type of person we are attracted to will change as we grow older and mature in our tastes. In our youth we may be looking for someone that is perfectly physically attractive in our eyes, and who also has the approval of our friends. As we grow older, we may be prepared to compromise our physical attractiveness standards, relaxing them into more realistic criteria, while looking for a partner who is our equal. Later in life, we may have been through a divorce or similar life experience. We have certainly gained experience and wisdom. At that time in life, we may give much more latitude in appearance and be more interested in stability and affection.

The First Glance

The very first impressions will be based on the person's physical appearance and their mannerisms. These would include clothing, hair, cleanliness, neatness, countenance, posture, facial features, weight, height, proportions, fitness, health, facial expressions, hand gestures, grace and voice tone. What we are looking for is their overall attractiveness.

What strikes you first about a person will change as you look from one individual to another. You might notice one person because of their hair, while another person draws your attention based on the way they move and smile.

From our own perspective, we think we know everything that we need to know about a person from that first glance. It is highly discriminating, unconscious and instantaneous. It is a natural, and actually an important, survival tool. Since the earliest days of mankind, it has been necessary to be able to instantly discern danger, and tell the difference between friend and foe.

> *It is important that we develop suitable discernment standards for each of the relationships we choose to develop and the people that we want to be associated with.*

Without these discernment standards, we would be attracted to everyone equally. Without our critical factors, it would be as natural to be in love with Hitler as it would be with Princess Diana. Of course, this type of discernment should not be confused with generalized prejudices.

So, we think that we know everything about a person from that first glance, but in truth we are probably quite off the mark. How many times have we dismissed a person or failed to be impressed until we got to know them better? How often have we been amazed at our depth of fascination in another person - but not until a lengthier conversation? It happens all the time.

However, that's not the point. The point is that we, as humans, won't normally get that far if the initial attraction isn't there first.

It is true that a lot of relationships begin after a person has had numerous contacts with a person over a period of time. In these cases, the person has had the opportunity to warm up to that other person based on something more than just the first glance. But we don't always have that opportunity. And, certainly, creating that scenario, and that opportunity takes time, patience, and perseverance - things that, in our modern world, we don't always have a lot of!

So once again, the initial impression - that first glance - is so vitally important.

Even if you do get the opportunity for future contact allowing for a deeper exploration of the personality and character of another person, any negative first impressions will have to be overcome.

Wouldn't it just be simpler to start out on the right foot, with a great first impression?

When I (ML) was about 11 years old, my father, Maurice LaBay, made a statement that had a profound impact on me - and still does. Interestingly, when I told him I was going to quote him on this, he didn't even remember making the statement but said that he still agrees with it. And also. keep in mind that I was, indeed, 11 years old. I was starting to develop, but still trying to lose baby fat. My face was generally broken out and I had incredible doubts about ever being beautiful or attractive - even though my loving parents kept assuring me that I was.

To set the scene, it was back in the days of airport technology when passengers, while embarking and disembarking from airplanes, had to walk down a set of stairs onto the tarmac. (Yes, I am that old!) Ann-Margaret was slated to come to Denver's Stapleton Airport, and somehow her flight arrival time was made known to the public. So my father took my brother and myself to the airport to get a look at Ann-Margaret. We had parked along an adjacent road and were standing there, in the windswept field, with our fingers locked onto the chain link fence that separated us from the landing strip. (Later, the Beatles would make this same entrance into the Mile High City.)

As she stepped out of the plane and waved to her awaiting fans, my father, nonchalantly, and obviously unconsciously, said, "Any woman can make herself attractive if she just fixes herself up."

> *Anyone is capable of being more attractive if they are willing to put effort into it.*

A simple concept, yet profound.

We aren't all going to look like Ann-Margaret, but we can all find our own style of beauty and enhance the natural features that we have.

There are almost endless ways that a person can be more attractive. Women seem to have the most options, yet both genders are capable of radical improvements if they put the effort into it.

Based on the list of sorting factors listed above, let's take a look at some of the things people can do to become more attractive at the first glance.

Clothing and Jewelry

We all have to wear something! Well, I suppose there are some exceptions. However, what you choose to wear tells a person a lot about your taste and your sense of self.

> *Since time immemorial, how we decorate our bodies has shown to the world our socio/economic status, our marital status, our occupation, and our activities.*

There are school uniforms, company/work uniforms, and sports uniforms. There are clothes that separate the blue-collar worker from the white-collar professional. Isn't it interesting that this socio-economical level is even referred to by their clothing? Teams will have their own clothing and colors, and gangs will have their trademark dress code. Our clothing shows a sense of belonging to a certain class or group.

In this same way, when you go out of your house, the clothing that you wear tells a person a tremendous amount of information. The impression doesn't have to be accurate - but it is the first impression. Many well-to-do people will go to the store in tattered work clothes because they need to buy something for the garden that they are working on. "Clothing does not the man make." However, in real life, when you are better dressed, you will receive more attention and get better service from retail stores, restaurants, and so forth. Clothing does, in fact, make a difference.

So what do you do when you don't know how to put the "right outfit" or the "right look" together? First of all, don't feel bad about it. Most people don't have that knowledge. We are so busy trying to make a living or take care of day-to-day responsibilities, it is not expected that we can also keep up with fashion, our personal colors, or which way the stripes need to go in order to look slimmer.

Some people have a knack for style, or have the time to read the fashion magazines. But you don't have to be a trendsetter just to dress in a manner that presents your best appearance.

Free Expert Help

Some department stores offer fashion consultants. And many of the finer stores have trained personnel who can help you put outfits together and accessorize properly. You may also find some smaller boutiques have an overall appeal to your sense of taste for the look that you want to achieve. Their sales personnel are more familiar with the present season's line, and how the clothes are to be worn and accessorized, than the customer could ever hope or want to be.

You might find one or more of the clerks that seem to have a good "look" themselves, and let them "dress you up". It can be really fun, and they may be able to expand your horizons of what you think looks good, bring you more up-to-date with your look, and suggest combinations that you would never have imagined.

Of course, they are trying to make a sale. However, in the nicer boutiques and department stores they are concerned about having a happy customer who will return to them because they feel and look good in their new clothes. When you receive compliments on your new look, you will return for more advice and items.

You can also take a friend along on your shopping spree. Just make sure the friend has a sense of taste! Consider all the friends and acquaintances that you have, and narrow them down to the one or two who seem to have a sense of style and to be able to put a "look" together. This could be a friend or relative of the same or the opposite sex. Get their opinion on your choices, or have them help make selections.

Both my (ML) brother and my son have far better taste and sense of style than my daughter or I could ever hope to have. We are always thrilled when they will agree to go shopping with us and lend us their wisdom and eye for fashion. They, in turn, enjoy the hunt for the right thing. My brother, in particular, has an innate nose for finding the best bargains while never settling for anything less than the best. Everyone should find such a person to help with his or her quest for the personal look!

Remember that what looks good on someone else may not look good at all on you. And the reverse is true. You may look stunning in an outfit that just wouldn't suit other people. Whatever it is, carry it off with an apparent comfort, and with flair.

Certain looks will be congruent for certain types of people and will look absolutely awkward on someone else. When you are in the right outfit, you will feel and look relaxed, comfortable and at ease.

Hair

A person's hair is frequently the first thing that others will notice. It can be one of the most important first glance features that you can emphasize. At the very minimum, hair should be clean, healthy and kept. I won't say combed, because there are very fashionable styles that are purposely windblown, spiked, chopped and tossed.

With hair, again, get opinions from others who have an obvious sense of style and a critical eye. It is just so hard to "see" ourselves sometimes. That's the reason, I think, that we are so appalled when we see pictures of ourselves. The camera catches details that we grossly wipe from view when we check ourselves out in a mirror. Alicia Silverstone's character in the movie "Clueless" had the right idea. She would take a Polaroid picture of herself before going out so she could see what she really looked like.

If you have worn the same hairstyle for more than a couple of years, it is certainly time to consider an update! It is a good idea to look through magazines to find new and interesting ways to wear your hair. Some salons offer computer-imaging services. They take a digital picture of you, and have a computer that will show you what you look like with various hairstyles and colors. This is great if you can afford it, and if you can find a salon with this type of technology.

However, there are highly experienced stylists available at the better salons that can help you with your look.

There are many factors that will need to be considered. Face shape, skin coloring, facial features, hair texture and thickness. Careful consideration should also be given to how much time and effort you are willing to devote to maintaining a style.

Some men will say that they like women's hair to be long - the longer the better. Yet other men prefer shorter hair on women, because long hair gets in the way during intimate moments. If you don't have a specific partner you are trying to please, you will do best choosing a style that suits your face, hair characteristics, and lifestyle.

Women with really long hair seem to be under the impression that it is some type of contest. Perhaps it is an idea left over from early childhood. The impression is that it is some kind of honor or mark of excellence that you have been able to grow your hair out longer than anyone else you know. However, at a certain point, especially on mature women, pure length takes over and style and attractiveness may begin to fade.

Once again, face shape, hair texture, and lifestyle should be the deciding factors rather than competition.

Just to point out how trends change and technology influences fashion, let's think back to a time not so long ago. Back in the 50's we didn't have hair blowers and curling irons. At that time there was an issue of whether women should leave the house in curlers. And if so, should they wear a scarf over them, or not. Boy, does that seem like ancient history now. And thank goodness!

82

> *With today's modern hair appliances and with a "do" that suits your lifestyle, we can have a presentable coiffure any time.*

All women agree that they would prefer men clip their hair short rather than combing long hair across a bald spot. Most women even prefer bald rather than a man wearing a toupee. A balding head may indicate an abundance of testosterone, and a cerebral personality. A balding gentleman gives us the impression of a thinking, analytical type. That is very appealing to many women, and yet men who are exactly that type may be hiding their very attractive nature by covering their bald spots. Why not be who you truly are and let the people who are attracted to those traits find you?

A completely clean-shaven, bald head can be very alluring as well as fashionable. Take, for instance, Yule Brenner, Patrick Stewart, Bruce Willis in some roles, and Louis Gossett, Jr. They have shown us that bald truly can be beautiful.

> *Attention to your hair can go a long way in presenting your most attractive look.*

Don't you find it mind boggling to see that people will attend an event such as a singles dance or mixer, with the assumed intention of meeting others, and yet show up with "bed-head"? Brad Pitt might get away with it, but anyone else will surely give a better impression if they take a minute to wet their hair and recomb it.

It is phenomenal how easily a hairstyle can add or subtract years from a person's apparent age. Youthful, healthy hair can help to make most people look ten years younger and much more attractive, while a limp, outdated style can make a person look tired and add unwanted years to their appearance.

Be Neat and Clean

A lot can be assumed about someone's personal neatness and hygiene, just from a precursory glance. The overall look of the hairstyle and clothing, the care of the shoes, and the way that articles are stored or carried in handbags and briefcases are all blatant indicators of how a person handles the rest of their lives, as well.

When a person looks disheveled and their shoes are scuffed, observers will get the impression that they don't pay attention to their personal appearance, and assume that their lives and homes are equally in disarray. Likewise, a person whose briefcase or handbag is neatly packed with organized items, will give the impression that they take better care of their belongings and are perhaps more painstaking in their personal affairs. Like any first impression, it could be inaccurate. However, we are discussing the impressions given, not the reality of the situation.

Personal cleanliness is an important and delicate issue. Sometimes things just happen, and we find ourselves in a situation where our deodorant has failed. Most people have probably experienced this at one time in their life or another. However, cleanliness is a vital factor in being attractive.

As a general guideline, it is best to bathe frequently, wash clothes frequently, and find a deodorant that works for one's body style and chemistry. Not all deodorants work equally well for different people, or in different situations. There are many products on the market that will aid females with their particular body odor issues.

If you have allergies, hay fever, a cold or generally have dulled olfactory senses, you may want to check with a friend to determine whether you are smelling your freshest. If you are having a hard time smelling other things, you may not be able to notice your own body odors.

The use of perfume has become less and less favored in certain social circles. People are recognizing that others can be allergic to certain perfumes. Some doctor's offices are requesting that their patients refrain from wearing perfumes into the office for this very reason. And it is easy to sympathize with anyone in an office or similar situation, who may be exposed to dozens of people or more on a daily basis, each wearing a different and possibly conflicting scent.

But it is just as disconcerting to be on a date and be inundated with a heavy perfume when you are in the car together, or to give someone a hug and then have their perfume or cologne lingering on your clothes the rest of the day.

Just a drop of a perfume fragrance is sufficient to attract someone closer. If the scent is too heavily applied it may actually work in the reverse!

Remember that not everyone will share the same taste in fragrance. And, in intimate settings, many people would rather enjoy the smell of freshly bathed skin and natural pheromones than heavily applied colognes and after-shave.

Physical Features

With the advent of cosmetic surgery, there is hardly anything that can't be changed about a person's facial features. Generally, such drastic measures are not necessary. However, when it is necessary, or even when it is simply desirable, with sufficient funds, modern technology can accommodate us.

There is rhinoplasty (the "nose job"), liposuction (surgical removal of body fat), implants (on more body parts than you might have guessed), and tummy tucks. There is the ability to remove acne scars, bunions, and unwanted hair. You know those little frown marks about the bridge of the nose that people get? Those can be avoided by having the underlying muscles cut, or by injection of a chemical that creates the inability for those muscles to scrunch up.

Your teeth can be whitened, or pearly onlays applied, to create a perfect smile. Gaps can be closed and braces can straighten even the most chaotic set of teeth.

> *Given enough money, a person could pretty much stylize the perfect body that they want to dwell in.*

Short of medical manipulation, there is always diet, exercise and staying active. An active lifestyle readily shows in a person's energy levels, moods, attitudes and body. With determination, you can turn most any body into a shapely attractive package.

My (ML) son, Quincy Miller, works out and is becoming a fitness trainer at the time of this publication. While visiting me in Washington, he urged me to begin a more rigorous workout program to avoid the signs of aging and raise my energy levels and chances of a longer life. He purchased a Muscle Media magazine and showed me the pictures of a contest called the EAS (Experimental And Applied Sciences, Inc.) Physique Transformation Challenge. Unbelievable! In 12 weeks, through vigilant diet, exercise and supplementation, a decidedly overweight couch potato can turn their body into a moderate-to-fit, muscular god or goddess! Naturally, this takes discipline and dedication, yet it is possible.

Those pictures took away any remaining excuses that I had. There is no reason why a human being, with a little bit of effort, cannot be in at least moderate physical condition. Being overweight detracts from the first impression that you give people - and overweight people are aware of that. When a person is overweight, they are less likely to go out and do certain things, they have lowered self-confidence in certain situations, and are more likely to be overlooked for jobs and relationships.

Aside from diet and exercise, there can be some serious emotional and mental roots to weight gain. These can be addressed very elegantly and successfully through hypnotherapy and hypnosis.

> *Whatever fitness or weight loss program you choose to initiate, it is wise to consult a physician to make sure you are not at risk of injuring yourself.*

Posture is another aspect of physical appearance that readily gives an impression. If you are slumped over, slouching, or shuffling your feet, or if your head hangs down forlornly, you give the impression of low self-esteem, lack of confidence, depression, and being dejected and rejected. This is not going to attract others to you - except maybe to offer you assistance or compassion!

Conversely, standing erect and tall with the head level (not "stuck up"), gives the impression that the person is confident, self-assured, optimistic, alert and healthy. These are traits that are universally seen as attractive and commendable. The person is more likely to receive respect and admiration when they appear to give the same to themselves.

Grace and poise can add greatly to a person's attractiveness. A woman who moves smoothly and elegantly is generally considered more feminine and attractive than if she walks with a heavy step, is clumsy, or has bold erratic hand movements.

A man also becomes more attractive when he exhibits grace and poise. It is the promise of chivalry and romance, elegance and gentleness. An entertaining look at the development of grace and poise in a man is demonstrated in the movie "Dirty Rotten Scoundrels" when Michael Caine's character teaches Steve Martin's character to be a European-style gentleman.

Grace and poise in either gender gives the impression of self-control, good manners, and being comfortable in one's body. These traits can be enhanced in a person through diligent practice, by being aware of the body's movements and the space around one, and by participating in classes such as dance, yoga, and Tai chi.

Make-up

Women have a significant edge on augmenting their natural beauty, and concealing their minor flaws. The multi-billion dollar cosmetic industry provides us with the most available and affordable way to highlight and diminish what traits nature provided us with. And although some men use a small amount of cosmetic products, the make-up industry, aside from Hollywood, really belongs to the domain of the female.

As with clothing, make-up trends change with the season and there are always trendy colors and techniques - and those, which have faded from fashion. It is hard to stay up on the latest trends, and so, is best left to the experts and those of our friends who have a special interest in keeping up with the changes.

Fortunately the cosmetic industry is competitive. It is easy to acquire coupons, free samples and free makeovers. The major department stores feature oasis' of beauty, where trained specialists will give free facials, beauty treatments, and makeovers. They will teach you to apply your make-up properly, and give you advice on the latest trends in cosmetics. For these services they hope that you will purchase some of their products. However, that is not mandatory. You can sign up on their customer lists and receive notices in the mail about special events, offers of discounts and free "gift with purchase" specials. When you purchase any product from these counters, you may request free samples of other products, which the clerks are generally very pleased to give away.

Skin care is an essential step in being attractive. This is a fact that many women, and even more men, overlook.   Large clogged pores, oily skin, wrinkles, blotches, sags and bags are all noticeable problems that could very possibly be easily taken care of. The advice is free. Why not take advantage of such a simple attraction enhancement?

Once the "canvas" is clean and prepared, the colors can be applied. Make-up, like so many things, needs to be tailored to an individual. Again, this is why having expert advice is so important.   There are warm colors and cool colors, there are daytime make-up techniques, and those geared for evenings out. There are so many choices and application variations that it can sometimes seem overwhelming. However, after a quick demonstration, it all becomes quite straightforward.

Enhancing your natural beauty - your eyes, your mouth, and your cheekbones - will make you feel better about yourself. And when you start to become aware that more people are commenting on how attractive you look, you will be glad that you have taken the time to place some simple strokes of color across your face!

Make-up doesn't have to be heavy - or make you look too "made up". In fact, daytime make-up, when applied well, is generally geared to give a natural, fresh look.

> *Make-up trends also differ between areas of the country and areas of the world! The thick black eye-liner that works well in New Jersey will appear very much out of place on the beaches of California, or in a small Iowa farming community.*

So wherever you live, it is easy to check with the clerks at the cosmetic counters for an update on the local make-up fashion trends. If you are traveling, or have recently moved to a new region of the world, be alert to how people around you are wearing their make-up. You can also give yourself a treat, and schedule a personal luxury hour at the local department store cosmetic counter. There is no better inexpensive way to update your look and learn to take better care of your skin.

Facial Expressions and Hand Gestures

Among the very first things that are noticed about a person are their facial expressions and hand gestures. These are "features" that truly are within a person's control.

A pleasant, relaxed countenance that responds to conversation with a natural smile speaks of alertness and interest. Sparkling eyes and an attentively tilted head are all looks that gain universal approval. A smile not only changes the way people respond to us, it changes our own internal chemistry.

> *When we smile, endorphins are released into our bodies, elevating our moods and energizing us.*

So a simple smile can initiate a whole chain reaction of positive feelings. Try a simple exercise to prove this to yourself:

> *Put a serious, yet relaxed expression on your face.*
> *Close your eyes.*
> *Notice how your body feels.*
> *What do you notice about your energy?*
> *Now smile broadly.*
> *What do you notice happening in your body?*
> *What changes do you sense occurring?*

Smiling not only changes our mood and our energy, it changes how we pronounce words and it alters our inflection or our "tone of voice". It makes our words sound more pleasant, and our smile is apt to give a lilt to our voices. We can readily sense when a person is smiling on the other end of a telephone. Why? From the way that the smile has altered the shape of our mouths and, therefore, our speech patterns and intonation.

Some people never crack a smile, look very serious, or frown a lot. And while there is a time and a place to be serious, if that is someone's typical mode of operation, it is quite likely that it is putting people off.

Jack Swaney, our illustrator, used to be my (ML) neighbor. We would go out for exercise and adventure pretty regularly, but the distinct habit we had was going to Starbuck's in the morning for lattes. We would walk there, since they are conveniently located to every resident in the greater Seattle area - no matter where you live!! It was our opportunity to people watch, read the newspaper and catch up on the gossip in each other's lives.

Every day, one young lady, probably in her twenties, would come in to get her "fix" too. She was proportionately built, fairly well dressed and would have had average to attractive features - except that she always had her forehead scrunched up in deep furrows. I couldn't tell if that was the look of concern, worry, anger, fear, concentration, or pain. I even wonder if she knew that she was doing it because she had that expression every single morning. Being a therapist, I'm trained and experienced in looking for the deeper, underlying issues that would cause that behavior. However, my first, human, gut reaction, was "stay away from her, don't talk to her". Her demeanor was one that made people feel it might not be pleasant to be around her. That may have been completely untrue.

How often do we do that? How often are our faces showing something to the world that may or may not be true or congruent with this moment? How often are our expressions repelling rather than attracting? There is nothing more revealing about our demeanor and gestures than seeing ourselves on a candid film.

> *Everyone should go to foreign countries to learn about other cultures and people - and everyone should watch films of himself or herself to obtain a reality check!*

There are so many simple and inexpensive ways to make ourselves more physically attractive to others. We simply have to observe ourselves and notice the reactions that we are receiving from others. If we see that something isn't working, then we must look for something else to try.

If it is important to us to become more attractive to others, then it will become a priority to discover what, in ourselves, needs to be altered to get the type of reaction that we seek.

FOUR

Flirting Makes the World Go 'Round

Imagine you are a man, standing at the subway station waiting for your train. You are scanning the headlines of the newspaper you hold in your hand. It's been a long day and you are feeling pretty drained. Visions of a hot shower and lounging on your couch are the only things keeping you moving towards home. You lean back against a support pillar and disinterestedly gaze around in anticipation of the train.

To your surprise you see a young, attractive woman looking at you. You look past her, but can't stop yourself from taking another look. Your eyes lock with hers and she gives you a broad smile that lights up her face. You smile back, somewhat in disbelief. As you smile, a tingle of energy fills your chest. You feel a little lighter, and you have forgotten how tired you were.

The train arrives and, having boarded, you find yourself standing right next to the smiling young lady. She looks up at you from the corner of her eyes, with a mischievous smile, and lets her body rock gently with the rhythm of the moving train.

She strikes up a conversation with you, her eyes flashing, pupils expanded and black, laughter in her voice. As the train comes in to the next station, the slowing of the train causes her to gently bump into you. She takes hold of your arm to balance herself, as she looks directly into your eyes.

You have long forgotten the troubles of the day and have fully regained your energy. Is it any wonder that you invite her to stop at a pub for a drink as you both exit the train?

Flirting is one of the most exciting activities that a man and woman can engage in. When we flirt, we are being highly complimentary to the other person. We are signaling to them that we approve and that we are, indeed, attracted to them. When someone flirts with us, we are flattered, and we can be truly energized.

> **Flirting is a way of acting for the purpose of attracting something or someone into our lives.**

So, in becoming irresistibly attractive, it is vital to perfect the art of flirtation.

Flirting is fun, it is contagious and it generates tangible energy. When you flirt with another person, there is electricity, a positive tension that is built up between you. When someone flirts with you, that energy can make you more alert, brighten your day, and radically change your mood. It certainly grabs your attention!

So how does one flirt?

Remember how simple it seemed in the "old days" when a woman simply had to drop her handkerchief or a man would lean against a lamppost and whistle as a woman walked by? Movies have made it look fairly simple and straightforward at times, but in the real world, the techniques may be a bit more complex.

The Physical Attributes of Flirting

It is interesting to note that when a person is alone or perhaps with others of the same sex, they might sit or stand in a more relaxed and casual way. However, when a member of the opposite sex arrives on the scene, both men and women will stand more erect, their stomachs pulled in, their shoulders pulled back a little more, and their heads held up a little higher.

> *There is an automatic, unconscious desire to appear younger, more physically fit, and more confident to members of the opposite sex.*

When in the company of members of the opposite sex, both men and women will engage in activities relating to preening. They will straighten their clothes and brush off imaginary flecks of dust. Men will adjust their collar, cuffs, or tie, and women will touch their jewelry or run their fingertips across the ends of their fingernails. Both men and women will fuss with their hair, smoothing, fluffing or patting it. Women are known to twirl strands of hair in their fingers, and to saucily flick their heads in an attempt to toss stray locks back away from their face and neck areas.

Both sexes may sit or stand with their legs slightly further apart, and place their hands on their hips or in their belts. This movement accentuates the genital area and draws attention to the shape of the hips and waist.

Men and women, alike, will gaze longer at a person they are attracted to, hoping to catch that person looking back at them. There are all sorts of flirtatious movements that can be accomplished with the eyes, which are dealt with in their own chapter earlier in the book.

A long, steady gaze gives the impression of honesty and openness. A twinkle in the eye further reveals interest, animation and delight in what the person is seeing.

The double take is another way for you to flirt with your eyes. To accomplish this, allow your gaze to rest on the other person for only a second or two before continuing to sweep the room with your eyes. Then look back at them a second time, locking eyes with them, drinking them in, and perhaps smiling slightly.

This movement is very alluring and compelling. It is complimentary as the person senses that you have found something notable and special about them. And you appear obviously interested. If that person is also interested, one of you should take this opportunity of heightened energy to make an initial contact with the other. An offer to buy a drink, a compliment on some attractive feature of the other person, or a question about recognizing the other person from somewhere could all be possible appropriate responses. There is a definite rise in the energy levels between two people during a double take that should not be left unattended.

A smile is an all-important essential in the art of flirtation. This subject has been dealt with more extensively in a previous chapter. However, at this time let us add that the smiles used in flirting may not be the same styles as ones used in appearing generally attractive. On occasion a broad, open smile is useful in flirting, but more likely a subtler smile will be the more demure.

A flirting smile-type expression can range from looks of a pout, bemused, amused, an up-curl of one side of the mouth, a playful smirk, a relaxed slightly opened mouth, a friendly smile, a humored smile and an open mouthed "ahhhh" smile. Using only a broad grin would make one appear too anxious, and could eventually take on the appearance of being plastered to your face.

> *Perfecting the flirtatious smile: It is highly recommended to spend some time in front of a mirror making faces at yourself!*

By looking at yourself in the mirror, you get an idea of what others see when they look at you. What we think we look like is often inaccurate and not what others are seeing.

Women have the advantage of additional modes of flirtation that men will find irresistible. They can fondle a wine glass stem or other item in a slow and seductive manner. They can touch their neck, or run their hand along their thigh, in an unconscious movement that is alluring. While drawing attention to those parts of their body, these movements also indicate a level of sensuality that is potentially available to the lucky man.

Mouths and lips are very sensual parts of the body, and women have the advantage of having many options to show theirs off with charm. They can apply lipstick to enhance the color, shape and fullness. Lips, by nature, get pinker and puffier when the person is aroused, so in applying lipstick, the female has the opportunity to exaggerate what nature would do anyway. Lip-gloss gives the lips a wet appearance. Wet and slightly opened lips give a marked sexual invitation.

It is very inviting to a man when a woman touches something to her lips. It could be a glass, a pencil, a straw, a finger. Anything that makes the lips pucker and demonstrate their soft suppleness.

Most men agree that they enjoy having a woman flirt with them first. It is extremely flattering and signals them that it is a "green light" for them to flirt back. Because men are generally seen as the aggressor in all that they do, they are faced with the prospect of rejection much more often. When a woman shows interest in him, the man has an opportunity to relax and allow himself to be more vulnerable.

Another reason why men enjoy having the woman be the initiator in flirtation is because it is often a delicate matter for men to discern how much he can flirt before it is seen as sexual harassment. In our society there has been such an outcry, and so many legal battles, over sexual harassment, that the act of flirting, for a man, can be a dangerous tightrope at times.

> *Women have rarely been rejected when they show interest in, or initiate contact with, a man. It is, in fact, generally most enthusiastically received.*

Women can also make themselves alluring by glancing at someone from over their shoulder, head tilted back. She can sit, her own legs entwined around each other, and leaning towards the person of her attention and desire. This gesture can be made even more compelling by allowing a shoe to slip partially off a foot, and gently rocking it on the tips of her toes, thrusting the foot in and out of it.

When sitting with a man, a woman can slowly cross and uncross her legs. This movement shows off her legs, rocks her hips and drives a man crazy as he observes her legs rubbing together.

Speaking with a soft, sensual voice is also magnetic. Not only does it give the listener the impression that the person speaking is gentle and approachable, but it draws them in as they come closer to hear the words. A soft voice for a man or for a woman is so much more sensually magnetic than a loud, boisterous bellow, or a high-pitched nasal shrill. Practice your sultry voice and become comfortable with it in the privacy of your own home. Later, it will feel more natural in social settings.

Remember that when you are in a place where the music is really loud, if you move in close to a person's ear, you can speak in a rather soft tone and still be heard. Too often someone will come up to another person and shout directly into their ear, thinking that because they can't hear over the music, the person won't be able to hear them. But with the proximity to the ear, a softer tone is sufficient.

Is there a more marvelous opportunity to move closer to an attractive person than finding something to say to them in a noisy location? Speak softly and watch them come closer to hear what you are saying. Speak even softer and they will have to offer you their ear, which is right along side of their neck and cheek. How easy to touch your cheek against theirs as you softly deliver your message.

At this point you might have the opportunity to also gently hold their arm to steady yourself so close to them. They will have the pleasure of noticing your oh-so-delicately-applied perfume or cologne. Or he may take the opportunity to brush her hair aside, touching its silken softness.

Leaning in towards another person is a sign of interest in that person. The distance between your bodies will be a gauge of the attractiveness you both feel for each other. The spacing and positioning will also indicate how comfortable you are with the other person, and within your own body and space. Refer to the information presented in this book on body language for a more in-depth study of positioning and spacing during conversations.

Rene Russo's character in the recent movie version of *The Thomas Crown Affair* gave a rousing example of an effective flirtation when she leaned her body in to Pierce Brosnan's character, speaking very softly and directly to him, eyes locked on his. Her slow, graceful movement forward created tangible electricity between them. Their eyes showed their mutual interest. Her body, more than her words, revealed her intent. And then suddenly, she broke away, leaving him frustrated and wanting more, much more. He was challenged, and the chase was on!

Always leave them wanting more!

It is much more alluring to leave while they are still at the peak of their interest in you, rather than to let the conversation or the activity fade to dull before departing. When you leave early, the other person will be remembering the moments with full appreciation and good feelings. Even when you are not the cause of the decline in excitement, if you stay with them past that point, they may feel that the electricity they felt with you is gone and is, perhaps, irretrievable.

It is much better to make your exit on an exciting note, leaving them hungering for more of your time and your energy. You want them to feel that they just can't get enough of you. And you probably secretly prefer people about whom you feel the same way.

Keep in mind that it is most appealing if the woman breaks away first. If the man leaves the woman, she may take that as a sign that he is not a reliable mate and may wander even in a relationship. However, when the woman chooses to leave the conversation first, the man sees this as a challenge. She makes herself out to be more rare and valuable, and not too available to him or anyone else.

It is never attractive to be seen as needy or co-dependent. Especially in the early stages of a relationship, indications of these traits puts undo pressure on the other party. They may quickly begin to feel trapped or weighted down with undue responsibility. Both men and women will be well advised to avoid being clingy and placing too much emphasis on being accepted by this one person.

Keep things light and playful, enjoying the space between you. That's where the electricity resides. The longing creates the magnetic force that draws a person closer to you.

And while a woman needs to create the air of challenge to the man, she also is very attracted to the man who will indeed chase her. Since ancient times, a woman has needed to have proof that a man, in fact, desires her, wants her, will cherish her, and will do whatever it takes to conquer her heart and make her his own.

For a woman, a man's show of persistence is very persuasive. By his strong level of interest and willingness to work for the relationship, she is convinced of his conviction and commitment to making it work over time. A man who she may not have had so much interest in at the onset of their developing relationship, may eventually become the most attractive suitor available to her in her eyes. His persistence, along with a show of his ability to provide for her, through his gifts and where he chooses to take her on dates, will convince her of her ability to trust him and rely on him for her safety and financial security.

There is, of course, a distinction between persistence and stalking or harassing. If a person is interested, yet is simply not readily available, they will continue to flirt and be friendly. If they have firmly said no, it is important that their wishes are respected.

When a person shows no interest, to further pursue them, or to go out of your way to reject them in return, is not particularly attractive. It is much better to remain friendly, showing your grace and dignity. Over time, their attraction to you may increase, and you can take up where you left off at a later date.

A man wants to know that he is desired and desirable, yet the challenge of the pursuit is all-important. The greater the challenge, within reason, in conquering his object of desire, the greater the victory!

For a woman to master that delicate balance of availability and challenge is the dance of courtship, indeed!

Flattery

Some say that flattery will get you nowhere. And others claim that flattery will get you everywhere. The fact is, we all love a good compliment.

If we are attracted to a person, and are flirting with them, it should be rather easy to find something to compliment them on. The key is to make your compliment personal and sincere. A woman loves to be told that she is beautiful, gorgeous, alluring, and attractive. And these compliments will go a long way. However, in the back of her head she may be thinking, "Oh, you probably say that to all the girls."

Be sure to intersperse these generic, albeit positive, compliments with comments that are specifically tailored for the person.

> *"Your hands are so delicate and soft."*
>
> *"The sound of your voice makes me melt."*
>
> *"Your poetry reading this evening had the audience captured. And you have certainly captured my attention."*

These statements show that the person giving the compliment is noticing something specific about the other person. An easy way to create a personalized compliment is to state what it is about the person that makes them attractive. Instead of saying, "You sure are attractive", you can state it as, "I immediately noticed how enticing your eyes are".

When complimenting another person, it is important to say it with sincerity. A comment can too easily be taken as a shallow pick-up line. It should be given as an open, true statement and then move on to another subject. As much as it is fun to receive a compliment, languishing on the subject could become embarrassing for the recipient, and make the delivery sound obsequious.

Create an Electric Current Between You

Flirtation is usually very charged with electricity. By the nature of the attraction, the unconscious body movements, the chemistry that occurs within the body, and the intentions in the mind, electricity is generated. All you have to do is direct it to that special someone.

It can be delivered in the initial look, through your eyes. It can be enhanced through your hands at the first handshake. See it, feel it and send it to that other person.

Keep the electricity flowing. It is exciting and energizing. It is the fuel that keeps any relationship going. And if your relationship develops, and becomes a long-term commitment, that electricity will keep the passion alive forever.

That very electricity will surround the two of you, excluding the rest of the world from the bubble that is your own private shared space.

Close out other people from view.

While you are manifesting that bubble of electricity, you are framing that person within your field of vision and in your field of energy. When you focus your attention on that person, anything else in that room, or anyone else, will fade from view. The world will simply melt away.

And if you have captured the attention of that other person, the world will fade away for them, as well. You will also become the center of their attention.

It can be a very magical experience to have that happen.

A friend of mine told me a story about someone who she found to be particularly good at making this type of magic occur.

"While at an event, I struck up a friendship with a particularly handsome and charming man. Whenever we were together I felt that he was only interested in me. During a party, while standing across from me in a group of people, he paid me a compliment about my outfit.

In that moment, the world faded from me. I didn't hear the din of chatter all around me; I didn't see anyone else in the room. Only his gaze locked with mine, and the sound of his words. Then the spell broke and all was as it had been. The room came back into view along with the animated conversation occurring all around me. What had occurred was like a personal slice of reality that

*only the two of us shared for that
instant."*

This is an art that can be perfected with practice. It
can be a profound experience shared by two people,
bringing them closer together into intimacy.

Reach Out and Touch

Reaching out to someone is very attractive in many
settings, and particularly in the act of flirting. As an
acquaintance enters a room, you might extend one arm
in their direction, and then beckon them to join you or
your group. That is a gracious invitation and gives a
warm feeling of being wanted and desired.

A man can reach out to assist a woman in any
number of situations. Perhaps she is getting out of a
car, coming down a flight of stairs, crossing a busy
street or walking on an icy sidewalk. He may also
reach out to assist her in being seated at a table.

A woman may reach out to a man in response to his
offer of assistance. And she also may use that type of
motion to touch him or gesture him in closer.

Whoever reaches out, and under whatever
circumstances, it is generally received as a sign of
welcome, acceptance and being connected.

Touching another person, however, can be a
"touchy situation"! Most people love to be touched in
some fashion, some of the time, by some people.
However, there are people who do not want to be
touched anywhere, by most anyone, at any time.

So, when meeting a person for the first time, if you
want to touch them, do it in small ways at the
beginning. By being sensitive to the other person's
reactions to your touch, you will be able to determine
whether this particular person is receptive or averse to

being touched. Some people enjoy being touched, but are extremely selective as to whom they wish to do the touching. So even if you see them being touched by someone else, you are best advised to test the waters in relation to you.

The desire to be touched varies by gender. Research has shown that men are much more open to being touched, especially by women. Any place above the waist is pretty much fair game most of the time for most males. Below the waist can be highly acceptable as well under certain circumstances, and when it is a woman doing the touching.

The statistics show that most women are comfortable, most of the time, being touched on the hands, arms, shoulders, head and knee areas. Under certain circumstances, women may find it acceptable to be touched on the legs and feet.

Women are probably more hesitant about being touched due to being guarded against the possibly misguided intentions of a male touching her. This may be particularly true if she has ever been through a negative experience when she felt she couldn't control the outcome of another person's touch.

However, with flirting, and especially in the initial stages, the hands, arms, shoulders and head are, in most instances, fair territory for touching a member of the opposite sex.

So how do we do that in a way that is alluring, yet not overdone? How can we touch without offending?

Start with the handshake, making it a pleasant experience. Remember to send your electric energy through your hand and right into the heart of that other person. The handshake should be accomplished by offering your full hand, not just the tips of your fingers. Also be sure to take their full hand into yours.

The amount of tension used during the squeeze of the handshake is a debatable issue and seems to be a detail that eludes some folks. Practice using a pressure that is solid and tangible; avoid the "limp fish" or the "cruncher" effect. Ask a couple of your friends to give you feedback on the feel of your handshake. Oftentimes, we are unaware of how very strong we really are, or how we are too extremely gentle.

> *Note that people wearing rings are particularly vulnerable to pain when the "cruncher" handshake is used. If you notice someone grimacing, that should be your first clue to loosen your grip!*

If you are feeling particularly flirtatious with another person, you can use both hands during a handshake for extra emphasis. While shaking their hand with your right hand, use your left hand fingers to lightly brush the top of their hand, or touch them on their forearm or elbow. This gives the moment an added air of connectedness.

Holding the person's hand for just a couple of extra seconds gives a signal to the person that you are interested in them and that you enjoy touching them. This extra moment also disrupts the natural unconscious action of shaking hands and makes the moment more consciously remembered.

> *By upsetting the normal "flow" of the handshake rhythm, you anchor the awareness and focus that they have on you.*

We will talk further about anchoring techniques in just a moment.

Holding Hands

Holding a person's hand is a very common and distinct form of flirting. The palms of our hands are very sensitive, and our fingers are our main tools for touching this world. So, to hold a person's hand is undeniably a statement that you want to touch them.

Some people find it easy and natural to take hold of another person's hand, while others are more reticent about it, letting the tension and nervousness build around that act. If, while talking, you have been occasionally touching the arm or shoulder of the other person, it might not be too difficult to touch their hand, or take theirs into yours.

A friend of mine (ML) told me about what he considered to be a very tantalizing move that a woman made on their first date. She pushed aside everything that was on the table between them. She then reached across with both hands, in an open gesture for him to put his hands in hers. His reaction to that was that she was interested in him, intent on connecting with him, wanted to touch him and wasn't going to let incidentals like silverware and salt and pepper shakers get in her way. He was very impressed and responsive to that scenario. Obviously it was highly complimentary to him, and a most successful flirtation.

Telling a person that you read palms, as a means of getting to hold their hand, can be cliché and considered as an old "pick-up line". However, if both parties are being playful and silly, and there is obvious attraction between you, they will be delighted to comply and allow you to tell them the good news about how they will be meeting an extremely attractive and interested member of the opposite sex - immediately!

The key to this technique is to keep it short, light and playful. Once you have their hand in yours you can go on to asking if their hands are like their mother's or their father's, and then the conversation can move to other topics as you continue to hold and caress their hand. Remember that asking to read their hand was only a ploy, and does not have to be drawn out. You might even want to ask them to take a turn reading your palm.

By this point, you should be quite aware of how comfortable they are with being touched. If the other party withdraws their hand, or hides them in a pocket or under a napkin, you know that they are not open to touching. If they playfully examine your hands too, or relax their hands in yours, you are being given permission to have at least this much of a connection with them.

Other forms of touching include brushing up against the other person "by accident" as you reach for something or move past them. You can touch their arm as you compliment them on the sweater they are wearing, or touch their hand as you admire a ring or bracelet. Reaching across the car to open the other party's door is an overt show of interest in touching and being close.

If you are sitting beside one another, a man and a woman might let their legs touch or rest against each other. It may be just a knee touching under the table, or it may be that you are close enough that the entire thigh is against that of the other person.

A person once told me (ML) that he found it very seductive once, when a woman slowly reached across to his face and brushed a crumb from his cheek, while they were dining. He said she could have pointed to it, or mentioned something, but the fact that she took her hand, and ever so gently touched his face, was what really felt seductive. It indicated to him that she cared about him and was interested in him.

As was mentioned before, you can anchor certain feelings by the use of touch. By anchoring we mean that when something is repeated, such as a touch, in a given context, the same mood or feelings can be regained by repeating that same touch. In other words, if you touch a person's elbow each time that you compliment them and make them feel better, later in a different situation, you can touch their elbow again giving them unconscious ties to those previous good feelings.

We see the effects of anchoring all the time. How many times do you hear a particular song and immediately remember some incident or event in the past? Our high school years are filled with such anchors. As are other important events. What memories are conjured by the following sensual experiences?

- The smell of bread baking
- A Beatles' song
- A police car with the lights flashing
- The scent of a Christmas tree

Studies show that scent is the most effective anchor. However, because music trends permeate our experiences, songs will certainly take us to memorable - and not so memorable - moments in our personal histories.

When using the anchoring touch, be certain that you are touching the person at a moment when they are having a positive experience with you. It is certainly not to your advantage to anchor bad or uncomfortable feelings.

So to anchor a positive feeling, you might consider the following scenario:

A man and a woman are standing side by side at a social event. They are engaged in a conversation. Each time he pays her a compliment he gently touches her elbow.

> *"I noticed how attractive you were and had to come over to say hello."*
> *(Touch)*
>
> *"You look so familiar, perhaps it is your beautiful eyes." (Touch)*
>
> *"That's a very interesting story, how long were you over in Europe?"*
> *(Touch)*
>
> *"I've really enjoyed your company. I hope I can see you again soon."*
> *(Touch)*

To her subconscious mind, that touch will now be directly related to how good she felt when he was talking with her. Perhaps that made her feel more open and trusting. So in the future, when she is touched in that same spot, she will once again feel more open and trusting.

Touch can also be used as a gauge of a person's willingness to get closer with another person. If a man touches a woman on the knee while making a point, for instance, and she doesn't stop him or move away, he can take that as a sign that she is comfortable with that level of touch. With further touches, her resistance begins to fade, and eventually and perhaps slowly, he can advance his touches to more and more intimate suggestions.

Go ahead and be bold. Take the initiative in the advance of flirtation. Someone has to do it, and you can't rely on anyone else to make it happen. Flirting makes everyone feel attractive, welcome and desirable. So how can you go wrong?

When starting out, don't take it really seriously. If you flirt lightly and in more subtle ways, if won't be a great loss if people don't flirt back. Many people have a flirtatious personality and flirt with others continuously whether or not anyone flirts back. There is nothing wrong with that. They are simply having some fun and making people feel better about themselves.

Of course we are making a distinction between flirting and making serious sexual overtures to another person. Flirting is light, playful, alluring and coy.

While flirting, only offer what you are willing to give. It isn't fair or kind to lead a person on, causing them to think that you are willing to go further sexually, or that you are more serious about a relationship, than you really are willing to live up to.

It takes practice to really master the art of flirtation. But it shouldn't be looked at as a drudging chore to learn. See it as one of the most exciting and rewarding past times a man and a woman can pursue. Flirting gives a person energy, a sense of being attractive, and is that which makes this world such an interesting place to live in!

FIVE

The Second Impression

We have discussed how we all want people to perceive us in the best light possible. In the initial moments of meeting someone, we are judged in a positive or negative light. In these first moments, if we are near the person we wish to attract, we can do a lot to help or hurt our cause.

The second moments happen after the initial attraction and saying "hello." If what happens next is taking place at a dance club, the experience will be very different from what we will be talking about in this chapter. Anytime you are in a specialized environment whether a dance club, a church social, a baseball game, a convention, or a Las Vegas showroom, you have an advantage in "first contact." In situations like these, you can meet people and talk with them about everything that is happening all around you. Meeting people and having them find you attractive in these specialized environments is enhanced because there are usually a lot of people and they all have the common interest of "being there." Everyone sees everyone else as a little bit more like they are, increasing the attractiveness of everyone at the event, even if it is a small increase.

Making a good second impression is a bit more difficult in non-specialized environments. In a restaurant or bar for example, the activities are not quite so exciting and unifying. In these settings you need to be a little more adept at putting your best foot forward.

Women and men, on average, have certain preferences in communication styles. For several years I have been researching the nuances of the public first meeting between two people. Men and women report some preferences as being similar and others as strikingly different in these first meetings.

Both men and women are influenced by the physical appearance of others when they first meet. One recent study revealed that on blind dates, both men and women would be interested to some degree in another date with a physically attractive person. Both men and women on the blind dates said they would be less interested in dating someone that was not physically attractive in their mind.

When developing rapport, it certainly is important to look as good as you can. We discuss this at length elsewhere in this book. Beyond physical appearance, we have learned that precisely where you sit or stand seems to impact how you are perceived.

In research I have done, men reveal that, on average, they are more comfortable when a woman is seated at a right angle to the man and not directly across from the man. Why this is the case is unclear. It is possible that men like to be in control of an environment and prefer to have a "clear view" of what is ahead, as much as possible.

Women, on the other hand, express that they prefer a man to sit directly across from them in contrast to having someone sit beside them or at a right angle! This contradiction in seating and standing preferences is one possible explanation for the difficulties men and women have in first contact situations.

In addition to this fascinating contrast in position preference, men and women (typically right-handed) who converse with people directly in front of them, almost universally prefer that person to stand or sit slightly to their right side in contrast to their left side. In other words, people prefer to communicate with each other when their right eyes are in alignment and not their left eyes.

It's possible that this is true because of left brain/right brain phenomena.

It seems that the left-brain is typically more dominant in language, rational communication and mathematics. It appears that the right brain is more immature, emotional and volatile than the left. When looking through one eye or the other, we are connecting with the opposite brain hemisphere.

As mentioned, there are six basic emotions that are identifiable: Fear, anger, sadness, disgust, interest and happiness. It is possible that when we first meet someone we tend to be uncomfortable. It's also possible that that discomfort may somehow relate to, or trigger, negative emotions that may tend to be more of a right brain phenomenon.

If this is indeed true, then when we look to our left, we would access more right brain activity than left and enhance the chances of becoming in touch with those more volatile and emotional experiences in memory. Could this be why people report more comfort when people are seated to their right hand side? It certainly hasn't been proven but the evidence mounts!

Therefore, when you meet someone new, you may want to play the percentages and meet people so they respond to you in the most favorable light possible. This means that when you are meeting someone, you can shake their hand (or participate in whatever greeting is appropriate to the occasion) and then keep that person to your right as you maintain eye contact with them.

Allow yourselves to be seated so that your right eyes are opposite of each other. This may just give you both a slightly more comfortable feeling than you would have in a different setting.

It's interesting to note that in public demonstrations of this phenomenon, people who sit to the left of another person tend to describe the "first emotion that comes to mind" as "fear" or "anger". When people sit to the right of another person, the person tends to describe their first emotion felt as that of "fine", "glad to see you" or "happy". The results do not prove the causality, but the evidence is strong. This leads me to believe that your best first impression comes when the other person is seated to your right!

What if the other person is left-handed?

It seems that about 70% of left-handers respond as most right-handers do. The other 30% seem to be indifferent or prefer the exact opposite preferences we have been discussing.

If all of this seems awfully technical and a lot to remember, just keep in mind that, **"right eyes lining up makes everything all right".**

Mirroring

There you are. There they are. You are seated across from each other. The right eyes line up. Now what?

As we discuss elsewhere in this book, like really does attract like. We all tend to like people who are like ourselves. From a physiological perspective, this means that the best impression you can make on another person is probably to emulate their physiology, their posture and even some of their non-verbal mannerisms.

When you are in that first contact situation, you can easily notice how the other person is seated and then model their physiology. This is done by adopting a similar body posture and seeing to it that your hand and leg positions are similar. In other words, simply sit as if they were looking in a mirror and seeing themselves. You will find this easy to do.

This process of mirroring another person is often called "matching" or "pacing". When people begin to use similar gestures and experience similar facial responses, they tend to synchronize in other ways that are almost magical. This mirroring can lead to rapport being developed more quickly and sometimes instantly.

Leading

It can be very useful in situations of attraction to know if your belief that you are synchronized with another person is correct. One strategy that you can use is to briefly stop mirroring the person you are with and do something different. Take a drink from a glass of water, move forward, smile, etc. See if they move out of the position that they were in. If they do, you probably are synchronized with the other person and are, indeed, in rapport. The other person doesn't have to match your exact behavior. You simply want to know if they move in some way.

If not, continue pacing (mirroring) until they will accept your lead into a different physiology.

The nonverbal dance of pacing and leading is one that will come with practice and when you have become more comfortable in first contact situations.

Pacing and Leading With Your Voice

The dance goes beyond non-verbal communication.

Have you noticed that people speak at different rates? Do you feel comfortable with people who speak at a very different rate than yourself? Most people don't. Imagine for just a moment that two people are having dinner and one of them is speaking at a mile a minute. Words flow from their mouth like water from a fire hydrant. The other person, on the other hand, speaks as if there is a tightly woven coffee filter in their mouth, slowing the pace of their communication down to a virtual standstill. These two people are not likely to build rapport with each other. They process and articulate information in a different fashion from each other.

It is generally easier for people who normally speak quickly to slow down than it is for people who speak slowly to speak more quickly. People who speak quickly normally perceive slower communicators to be dull and boring on the negative side and gentle and caring on the positive side. The fast-paced communicator often becomes impatient with the pace of the slower communicator.

People who speak slowly consider those who speak quickly to be aggressive and often rude on the negative side. Positive attributes ascribed to the fast-paced communicator include intelligence and quick thinkers.

We believe there are approximately three different "speeds" at which people communicate. Each of these speeds will change throughout a conversation depending upon the content of the conversation. Depending on a person's default speaking pace and their mood, a person will speak in one of three ranges of pace: fast-paced, moderate-paced and slow-paced.

People who speak quickly are generally visually-oriented and speak as quickly as they see the pictures in their minds. Medium paced communicators seem to be very auditory in nature. In other words they tend to speak in such a manner that points to the quality of the words they say. They may hear their own voice more clearly than other people. They may have a greater awareness of the affect their voice has on other people. Finally, the emotional communicator is the person who tends to speak more slowly. They seem to sort their thoughts through their emotions. They might be more sensitive to people and therefore careful about what they say.

If you can pace the other person's speaking style in a first contact meeting, your partner will see you in a more positive light. As was noted earlier, it is easier to slow down in communication than it is to speed up! Regardless, moving your "speed control" in the general direction of your first contact partner will enhance their impressions of you as to being more like them. This increases rapport and makes everything that happens for the rest of the meeting more likely to succeed.

We all tend to evaluate other people by the sound of their voice, the words they choose to communicate with, their tonality, and especially their pacing. Until you become unconsciously adept at matching another person's vocal pacing, consciously alter your own speaking habits so that they more closely match those around you.

How Can You Increase Your Pacing if you are an Emotional Communicator?

As we noted earlier, increasing your pace is more difficult than reducing your pace!

It is difficult but do-able. Begin practicing by reading a page from a book at your regular pace. Count how many words per minute you speak. Now, read the page again into a tape recorder and be certain to articulate all the words clearly as you speak more quickly. Do it again and see if you can talk just a little more quickly, maybe adding 20 words per minute. The idea is not to race through the page but to learn how to communicate more quickly. About 1/3 of all communicators are fast-paced, visual speakers. They will perceive you as more credible if you can speak at their level, or at least a bit more swiftly.

Kicking the Boredom Blues!

Have you ever been around someone who just talked and talked and talked and talked and...well...you get the idea! They droned on forever without ever coming to the point...or any point. Now, is it possible that this person was once you? We've all bored other people at one time or another. At a first contact meeting, we don't have the luxury of doing this! Therefore, here is an exercise that will help you kick the "I can be boring blues!"

Do this exercise when you are alone.

---

1) Think about a book, movie, or TV show that you recently read or saw.

2) Take two minutes to describe the show or book, out loud.

3) Pause, compose your thoughts and then do the same thing again in one minute!

4) Pause, compose your thoughts and do your review in 30 seconds!!

The ability to be brief and offer a "Reader's Digest Condensed Version" of any story, especially your personal stories, can be the most profound communication change you may ever make. By making your stories shorter, you give the listener the opportunity to have you elaborate on what you have just shared with them. If they don't ask for more, you may have just offered the right amount of information!

People who tell long stories while the rest of the group experiences eye-glaze are never perceived as attractive and are avoided. This simple exercise, done regularly, makes you instantly more attractive!

What Do I *Talk* About?

First contact situations tend to be a bit uncomfortable because you are now seated across from someone who you don't know that much about...

The good news is that you don't know much about this person. That really IS good news. You can explore every avenue and have a lifetime of stories behind every door. The opportunities are fraught with risks, of course. You could easily open a door for which the information you receive will be uncomfortable for one or both of you.

Most people ask about what a person's job is. That can be a good thing. Most everyone has a job. Research has shown that self-esteem is directly correlated to job satisfaction. If your first contact partner appears to have great self-esteem, this is a great place to go. However, do most people like their jobs? I don't know. You can ask about their family, but are most people happy with the relationships in their family? Maybe, but families and the events that happen within a family are a leading cause of distress to most people.

Therefore I like to direct the conversation just a bit.

Talking about the future can be nice, but there are problems with the future too. What if you find out a person is 45 years old and has decided that they have just spent 20 years in the wrong career. They don't want to go back to school, they have kids to take care of and feel stuck! The future may look more like the past. I avoid the future until a little later.

There are some directions that you can go that are almost failsafe, however. How do you like this?

> *"So, the end of the week comes. If money were no object, how would you like to spend your days off?"*

It seems to me that most people enjoy having an occasional day off. It also seems to me that most people would like talking about something that they can do if money is no object. I also learn a great deal about a person if I know what they would do with their time off. It tells me about what is really important to them in their life.

If you are meeting your first contact partner somewhere you have never been before, you can ask, "I've never been here. Have you?"

The answer is "yes" or "no." If they have, you can ask for recommendations as to food or entertainment. If they have not been to this place previously, then you can suggest an idea that might work.

Most environments have a theme of some kind. If you are in a restaurant, it may be Italian, or Indian or Vegetarian. Whatever that theme is, it provides you with an opportunity to discover more about your first contact partner. Have they ever been to Italy? To India? Have they ever considering becoming a vegetarian?

These are the opportunities to explore just a bit. If they have been abroad or have experienced an interesting lifestyle, this is your opportunity to get them happily lost in re-counting a positive experience, or, any experience that will begin a long and mutual self-disclosure.

Self Disclosure: Being Vulnerable, Not Unbearable

What is the right mix of conversation? How do you know what to talk about, what not to talk about, and how much? What kind of conversation is most likely to enhance your level of charisma and bring out your best you?

Most men and women want to discover what the person they are attracted to is really like inside. Everyone knows that there is more to a person than the "package" they are wrapped up in. To optimize attraction, we want to start off on the right foot. In the case of what to talk about, this means leaving negativity behind at first contact and revealing all that is good about you, your "personal attraction resume" and your experiences.

Self-disclosure means that you are going to share some of who you are with another person. Being vulnerable means that you can share a weakness or two but not necessarily weaknesses that are perceived as negative. (I might share that I don't like to change the oil in my car or I might let someone know that I'm not much of a handyman in contrast to my brother who is a carpenter.) If someone were looking for a handyman or a car mechanic, then I would be the wrong person to invest his or her time in. Regardless of what I might get out of a relationship, I'm not tightening and loosening bolts for anyone! So in a situation of attraction, just what are those areas that you are weak in, that you simply aren't going to change in? These are the areas in which you can be a bit self-effacing when communicating with your first contact partner. This self-disclosure makes your "personal attraction resume" more real. You've now disclosed that you are not perfect and have given them some of your positive attributes. First contact partners generally perceive this experience as positive. How might this conversation take place if I were speaking to a first contact partner?

Kevin: Nice to meet you. I'm glad you could have lunch today.

FCP: Thanks. This is a very nice place.

Kevin: Have you ever been here?

FCP: Never. Have you?

Kevin: A couple of times. I think you'll like it. When you go out, where do you like to go?

FCP: You mean restaurants?

Kevin: In general. Restaurants, movies, entertainment...

FCP: Hmmm...I guess I like just about everything...I like to go to movies...I like to dance and I like to take walks around the lake.

Kevin: What have you seen lately that you liked at the movies?

FCP: I thought American Beauty was great. Did you see that?

Kevin: No, I heard it was great. Without giving the ending away, what was it about?

FCP: Well, let's see. It had Annette Bening and Kevin Spacey in it and it's about how Spacey sort of drops out of the rat race and sort of takes time to smell the roses...(laughs) especially the ones he sees swirling around the girl who played the cheerleader in the movie.

Kevin: Thumbs up or down?

FCP: Definitely a thumbs up.

Kevin: OK, I'll give you another chance to play movie critic. What other movie have you seen lately that you thought was great?

FCP: The Sixth Sense was incredible.

Kevin: That I saw.

FCP: Did you figure it out before the end of the movie? I mean...about Bruce Willis?

Kevin: That he was dead? Yeah. In fact, I said it out loud and people turned around and were just a bit annoyed with me!

FCP: How did you guess?

Kevin: Because he never was touched by anyone. No one ever saw him except the boy. It was just a lucky guess.

FCP: I thought it was great.

Kevin: Me too. So, that's two thumbs up. (smiles)

(Waiter arrives)

Waiter: Would you like to order now?

Kevin: Not yet, can we have just a couple of minutes?

Waiter: Certainly.

Kevin: Some of the dishes are excellent. What kind of food you like?

FCP: I like just about everything. I'm pretty easy to please.

What about you?

Kevin: I'm a little more difficult to please than you are, which is why I wanted to meet you here. I don't eat most kinds of seafood, but the chicken, steak and pastas here are great.

FCP: You don't like seafood?

Kevin: I was raised in a family that ate a very Jewish-like diet. I still don't eat some things like crab and lobster. Some of my friends say I missed the boat. I tell them, it was a deep-sea fishing boat and I wasn't allowed to get on in the first place!

FCP: (smiles) Do you recommend anything?

Kevin: The Chicken Milan is pretty good if you don't like spicy food. If you do, you might want to....

And so it goes. The first contact partner and I are rolling right along. I disclose a little bit about myself and that will encourage the first contact partner to feel more open to disclose more about herself later on. There are numerous ways to deal with first contact. Wherever the other person leads you is where you want to go with conversation. Remember, everything is new in first contact so you have a wealth of life experiences to go to. The key is keep it positive.

What might a conversation be like that is not so effective in building rapport? Let's use the same scenario and just change the conversation a bit. Let's see how these people can ruin the first contact.

Kevin: Nice to meet you. I'm glad you could have lunch today.

FCP: Thanks, it's a very nice place.

Kevin: Have you ever been here?

FCP: Never, have you?

Kevin: A few times. The food is great. [Never promise something you can't fulfill.] Don't you just love steak houses? [Don't guess what they might like. Ask them their preferences instead!]

FCP: Not really, I'm more of a seafood fan.

Kevin: I don't eat seafood. I used to be a Seventh Day Adventist and it's just sort of stuck over the years. *[The church down the street from yours is probably a cult in someone else's mind, so leave the brand off the building!]*
FCP: Isn't that a cult?
Kevin: No, just a conservative church. They eat a Jewish diet, pretty much.
FCP: Huh.
Kevin: Hey, did you see Saving Private Ryan? *[How does this follow from eating?]*
FCP: Yeah.
Kevin: Great movie wasn't it? Most realistic war movie I've ever seen.
FCP: I thought it was gross. *[Time to skip this, Kevin, and go somewhere else.]*
Kevin: Well sure, that was part of the point right? Wasn't Tom Hanks brilliant?
FCP: He always is. I liked him in You've Got Mail. Did you like that?
Kevin: Yeah that was a good show, too. *[Why didn't you ask her what she liked best about it? Then you learn something about her and what she likes!]*
FCP: Very romantic.

Are these two people in sync? No. The reason is that neither of them is eliciting preferences of the other person. Instead they are making statements about their personal experience without thought of how the other person might feel about them. Too much disclosure, too fast and it isn't even significant disclosure. Each of these two people is in their own little world and neither is interested in penetrating the other person's reality!

The differences are subtle, but make all the difference in the world in the rapport-building process. This conversation is uncomfortable and going nowhere.

It's hard to be interested in someone who can only start a sentence with the word, "I." People who are irresistible are those people who are able to make other people feel important and cared about. When people start too many sentences with the word "I", the message is clear...you are interested in yourself. What follows are common mistakes people make in communicating, followed by the better way to express the thought, maximizing your irresistibility.

Mistake: *"I really like football. I'm a big Vikings fan."*
Better: "Do you like sports?"
[Wait for the response before delving any further. If you get a negative response, you've saved the person the struggle of being bored at a first meeting.]

Mistake: *"I come here all the time."*
Better: "Have you ever been here before?"
[So, you bring all the guys/girls here. That makes me feel really special. Thanks for making me just one of the many, instead of making first contact something special.]

Mistake: *"I just bought a '69 Chevy."*
[When was the last time you met a woman who cared or knew the difference between a '69 and a '79, or even cared? They are out there, but only at the car rallies.]

Mistake: *"My old boyfriend used to take me here."*
Better: "Yes, I've been here before. I really like it. I'm glad you chose it."

[No man on earth is interested in your past boyfriends or husbands. If the earth quakes and he happens to ask about former times and loves, touch on them and leave the subject. These are areas you should leave alone in first contact.]

Mistake: *"You have a beautiful body."*
Better: "I really like your hair."
[First contact is no time to be talking about her body. Your eyes should be focused from the neck up. Leave the rest for another day.]

Mistake: *"I love Mariah Carey."*
Better: "Who is your favorite recording artist?"
[Ask whom they find exciting first and see what it is about that performer that they like. Then share your interests.]

All of these simple but important changes in the way we communicate make us irresistible because we are focusing on the other person first. We can talk about our interests later. Learn what you can about the person you are with. The idea is not to agree that you like Mariah Carey if you don't, but to learn about the interests and loves of your first contact partner. You can detail why Tina Turner is your favorite female entertainer at a later date. Find out what they like and you can make your second meeting custom-made especially for your first contact partner.

Can't I Tell Them *Anything* About Me?

You can and you must participate in mutual self-disclosure so your first contact partner can learn about you as well. If you are too mysterious, then they will ultimately lose interest in you. However, most people

aren't too mysterious. Most people are too talkative.
Once on a roll, people tend to disclose far too much, far
too quickly. Even if your bank account has 7 digits, she
probably doesn't want to know yet. She'd probably like
to figure it out over a few dates. The best policy is to
always put your attention on the other person, their
interests, likes and dislikes. When you are asked about
your specific likes and dislikes, you should always be
frank and upfront. However, all things considered,
participate in less "I" and more "you". People will find
you more attractive!

When you are in a position of disclosure, you should
focus toward the positive. She may ask whether you
like your job. You probably hate your job. You can't lie.
You can be just a bit evasive, a bit optimistic and put
your best foot forward. Here are some examples. Your
first contact partner asks you the following questions.
Your first response right below the question is, of
course, true - but bluntly honest. In the second
response you are honest, optimistic and put everything
into a positive light.

"Do you like children?"

Poor: *"Kids drive me nuts."* [She probably has
               one, don't you think?]
Better: *"Isn't everyone a child inside?"*

"Do you like your job?"

Poor: *"No, I can't wait to leave."* [Sign of
               instability.]
Better: *"You know, I've been there three years. I
               have learned so much about the structure
               of companies. I can see myself there for*

three more years or maybe even finding something that would be even more fun and exciting. How about you?"

"Do you like wrestling?"

Poor: *"It's a stupid form of entertainment."* [Her brother is a wrestler, right? Whom did you just insult that is related to her?]

Better: *"I've never been to a wrestling match. Have you?"*

"Who are you voting for in the elections?"

Poor: *"Straight democratic ticket."* [50-50 chance you just offended him.]

Better: *"I need to know just a little more about the candidates. I like a lot about several of the candidates. How about you?"*

"What do you think about the company's decision to lay off 2,000 workers?"

Poor: *"They got laid off because they weren't producing enough income to justify their salary."* [Read that as your brother is a loser.]

Better: *"My hope is that people will find something that makes them happy and excited about their lives. It's always frustrating when people lose their jobs, but I bet you can remember times when people have told you it turned out to be the best thing that ever could have happened. You know what I mean?"*

"Do you think people should date more than one person at a time?

Poor:	*"Absolutely."* [50-50 chance that you blew their belief or attitude out of the water.]
Better:	*"Isn't that something that two people need to carefully evaluate? I can see problems if people don't agree on something as important as that, can't you?"*

"Where do you stand on the abortion issue?"

Poor:	*Any response.* [Deadly question if the two of you disagree.]
Better:	Smile and say, *"Wow, now I know how a presidential candidate feels. That is a really hard question to answer. You sit in front of someone you think is really incredible and you wouldn't do anything to offend them in any way. I think I would do whatever it would take to not answer the question. At least not tonight."*

In each situation, you could have said exactly what you thought. Unfortunately, in every case noted on the previous pages, that would have potentially polarized you from your first contact partner. The "better" responses are all respectful, optimistic and allow for a slower disclosure. They may think that it is completely reasonable to ask each of these questions, but few people think beforehand about what will happen if you completely disagree on something that is significant. Therefore, you need to be aware that a positive and optimistic response that allows for eventual and not immediate disclosure can be very wise.

How Will the Second Impression "Go?"

If the man believes the woman is relatively attractive and that the woman likes him, the meeting will go well from his perspective. He won't need to be impressed by much else. If the woman senses that the man has some status in his peer group, resources or some "potential", she will be happy with her first contact and consider a second. In addition to meeting at least one of these criteria, the woman will need to perceive the man as attractive to some degree to want a second contact.

139

SIX

The Secrets of Charisma

The Secret Ingredient of Charismatic Communication:

> *Make other people right...even when they are wrong.*

Now we can finally identify the magic ingredient, the secret recipe for charisma and the ingredients of irresistible attraction! What is it that makes us feel good when we are around certain people? How is it that some people make us feel very good about ourselves when we are near them? So good, in fact that we are drawn irresistibly to them, wanting more. You might think that it is something they are born with or something that they just come by naturally. You may think it is because they just clicked with you. More likely it is a result of the way they craft their words, the honesty and integrity in their communication, and the intent that rests at the very core of their thought. When the integrity is there, the words of appreciation, affection, and acceptance flow freely. Here is what I mean.

Imagine that you are in a discussion with someone and you are having trouble getting your point across. You have explained your point well, but your recipient is still feeling lost or confused. You have a choice of making them feel wrong for not understanding or making them feel good about themselves, even though they may be feeling confused and inadequate.

You might say, "I don't know why you don't understand, let me try and explain it in a way that you can grasp". In that communication, you have an intent that makes them feel wrong and leaves them feeling even more confused. Even if they are clearly unfocused and they weren't listening, it is still not going to help your position to make them feel bad for not paying attention. You can take this same dilemma and turn it around to your advantage. You say, "Maybe I am not explaining it well", or, "let me make it more clear". Now, you have shifted the burden to yourself to explain it in a way that makes no one wrong.

Your intention will work for or against you. If your intent is to make someone feel small, stupid, confused, or less than you in any way, you lose and so does the other person. You may think you emerge victorious when you have one-upped the other person, but in reality you have alienated them and linked some bad feelings to being with you. The more you do this, the more you build a state in that person that is associated with feeling bad. When you put yourself in a superior position it means that someone has to feel inferior in order for you to feel superior. If a person continually feels inferior around you, they are not going to have any desire to be near you.

On the other hand, placing a person at ease, helping them feel confident and appreciated when they are around is the magic of great communication. When you phrase a sentence by placing the burden on yourself, you take the pressure off of them and, more importantly, you remove any blame or need to make them wrong. This is where your intent becomes very important.

> *When you are about to enter into an important business conversation or presentation, or when you are enjoying a casual or romantic conversation, set your intention before you begin.*

Your words may not be clearly understood, your listener might be distracted or uninterested, you may be getting frustrated, yet the next thing you say brings your intention right back to center. "You are a very astute business woman and I can see by your success that you grasp concepts easily. I think I can explain it better like this..." Now your listener feels that you are acknowledging her sharp mind, her ability to understand and you have placed her in a position of wanting to pay attention.

Helping Your Listener Be Right, Even if He is Wrong!

Look at these next opening phrases and "make them right" patterns that are designed to help make your listener right.

"You are obviously a man/woman who..."

...demands the best out of employees
...appreciates quality
...enjoys the finest
...is very astute
...understands how others feel
...catches on quickly
...can spot an opportunity

"One thing I've noticed about you is..."

...how insightful you are about others
...how comfortable you are with yourself
...the way you inspire confidence in others
...how quickly you grasp the real meaning of a conversation

"I can tell that it is important to you that..."

...others see your point of view
...you demand the best in others
...you feel right about making a decision before you jump into it
...you make the most of your leisure time
...others are honest with you

"The thing I've always admired about you is..."

...your focus and vision
...how strongly you feel about your decisions
...how direct your communication is
...the way you always say exactly what you mean
...the way you embrace your dreams and hopes

The beginnings of these phrases can help to defuse a difficult moment, shift the intention of the person you are speaking to, make them feel right and shift their focus inward, no matter what they have said. It also gives you something to respond with when you are faced with a difficult person whom you need to have rapport with. And when you are not with a difficult person, these phrases will enhance rapport and help your listener to feel better around you.

When you are in conversation and you feel the need to be right, take a moment to catch yourself. If you feel this need, it surely means that you will be right at a cost. The high price could be the state of mind of your friend, boss, lover, or child. If you are right, then it can only make your listener wrong. And if your listener is feeling wrong, they are going to feel bad in your presence. And when they feel bad in your presence you will be very far from being attractive to them. So, resist the temptation to make others wrong, even if you are doing it subtly. Being a charismatic communicator means that others feel better about themselves when they are with you. It means that others look forward to being with you because they like themselves better as a result of being around you. The way others feel around you is often a result of the words you choose and your intention behind those words.

There are many ways to make other people wrong. You may see yourself in these scenarios and realize that in a subtle way, *you* have been making others feel wrong or bad around you. Some of these scenarios make others feel tense around you. Some raise suspicions. All of them create a negative feeling in your listener and will force them *to associate that bad feeling with being with you.*

On the other hand, making people feel right, even when they're wrong is just as easy and far more productive isn't?

One of the most charismatic entertainers of the 20th century was Elvis Presley. I'll never forget watching a TV interview with Elvis when I was a kid. I didn't know it then, but he never liked war protesters. He served his country and believed that everyone should. He never shared his opinions on issues like this, though. In the news interview, a woman asked, "What's your opinion of war protesters?"

"I'd just as soon keep my personal views to myself, I'm just an entertainer."

"Do you think other entertainers should keep their views to themselves?" (Referring to Presley fan, John Lennon, and other peace advocates.)

"No."

It was never Presley's place to make someone else wrong. He wasn't a learned man, nor was he eloquent. He stuttered and was very uncomfortable in front of reporters. He did have one trait that was worthy of note as a communicator: He didn't like to make people wrong.

Knowing what is important to other people is another one of the keys to fundamentally fabulous communication and irresistible attraction. Therefore when we talk with people and fail to learn about what is important to them, problems will arise!

Eight Types of Highly Unattractive Communicators

Have you ever been in a conversation where you found your mind drifting, dreaming, and struggling to stay focused? Do you remember how it feels to try and listen as someone drones on and on? When we are faced with a poor communicator, there can be many reasons for the missed connection. Often there are words and phrases that simply shut us down, and prevent us from listening as well as we would like. Many times, the person communicating is injecting so many negative words and ideas that we begin to feel down and heavy inside. It may just be that the person you are communicating with is boring you because the content of the communication is all about them, about stories you don't care to listen to, and people you have never met!

What if that poor communicator who is boring people to tears.... is *you*? How would you know if *you* are the one who is inserting negative associations, bringing up insignificant details, droning on about you, you, you? How do you know if someone is really interested in what you have to say... that they are really engaged in the conversation? What is your method of observing whether or not the person or group is interested and intrigued, or tired and looking for the door?

When you become a top-notch communicator, you learn from everyone you talk with. You will notice the subtle cues that tell you if you are in good rapport, speaking in a way that your audience understands, and using words that create desire and interest. You will be willing to identify in yourself those things that push others away and prevent them from listening as well as you would like. This is a very potent aspect of self-awareness that allows you to stay fascinating to everyone around you!

In this next section we will look at the areas of communication where people most commonly fail. You will discover how you may have been alienating others and helping them to feel negative when they are around you. As you read these scenarios, notice if you see yourself in them. Take time to be very honest about your style of communication and the effects you are having on those around you.

The Argumentative Communicator

Do you enjoy playing the devil's advocate? Are you constantly offering your opposing opinion when it is not asked for? Do you find yourself saying the word "but" often in your conversation with others? You may be an argumentative talker. There is an effective way to take an opposing view, but it may destroy rapport. There is a way to give your opinion, but it may be received as unwanted advice. When you continue to oppose the comments of your listener, you run the risk of making them feel wrong, stupid, or uninformed.

The Comparison Maker

Comparison happens when I share a thought of a feeling with a friend, and it might be something that is very personal, or something that I am looking for understanding about. The friend will offer up a response that tells me that she does not really care about what I have to say.

It might go like this, "I have been talking with my boss about how to handle these negotiations with Sally. I tried to get in to see him yesterday and he acted like he didn't want to talk to me about it."

Friend responds, "I know just what you mean! I had a boss once who was always finding time for everyone else, and every time I tried to ask a question she would brush me off. Once when George was talking to her, he... blah, blah, blah."

If you find yourself always looking to compare an event in your life with your friend, you can now change this nasty habit and develop the skills of the great communicators!

The "Better Than" Talker

The difference between a talker and a communicator is clear. The communicator is making an effort at understanding. A talker is someone who rambles endlessly without the intention of both people benefiting from the conversation. The "Better Than Talker" is very similar to the Comparison talker, but with a more condescending tone. The Better Than Talker is not comparing for purposes of being compassionate, but for the purpose of creating superiority. They are interested in feeling superior to the person they are speaking to and it requires that the listener become inferior. If the listener is feeling inferior, the talker is not in rapport, and any hope for a connection is lost.

The "Hear My Old Baggage" Communicator

There is a need for sympathy in some people that begs for pity. It may come out of a need to be rescued, or it may be a real cry for help. If you recognize this in yourself, take a look at why you need sympathy from others and why it is important for others to feel pity for you. Maybe you have had a very sad life and you really feel that you deserve a little sympathy. That certainly isn't unreasonable. Maybe you have had the short end of the stick and you feel you really have been a victim of some terribly unfortunate events. That's OK too. People DO have these experiences. The appropriate place to take these challenges is to a qualified therapist and work through your difficulties with them.

With the exception of recent events that demand sharing sympathy (someone losing a job or the death of a loved one, for example) there is little place for running through old baggage in conversation. Listening to old baggage by others places an obligation on your listener to feel something that they may not want to feel. It also connects being near you with feelings of sadness, need and despair. The more you do this, the more that others get those feelings connected to being near you. If you want to help others to feel bad around you, then you should try to get as much pity from them as possible. If you want others to seek you out and feel good around you, then you will want to save the truly difficult experiences for your trained therapist. They can listen with empathy and objectivity that friends and business associates simply aren't normally ready for.

The Judgmental Communicator

When Jason says, "Jim is really getting stressed. He must have some difficult clients right now", it is not a judgment, it is an observation. That's good.

When Cathy responds and says, "I know what you mean, he has never handled stress well. When he blew up at Ken the other day, he was so rude. He can't control himself and I am really tired of his attitude," that is a judgment. Cathy makes a statement of opinion as to what kind of person he is and how he is wrong for being that way.

If you judge others, you may think that you are doing it to gain rapport or be on their side. However, you may alienate yourself by showing your lack of self-respect. If you are not internally well aligned with yourself, you may find that you have a need to judge others in order to feel better than them.

Speaking judgmentally is a dead give away to others that you have issues of incompetence and insecurity. Don't play into it. Respond in a way that strengthens your position of self-respect and self-esteem.

In our example, Jason should probably respond back with, "Jim has always been helpful to me. I've learned a lot from him. He has his challenges, like we all do. Maybe he just needs a hand right now".

The Interrupting Communicator

The single most powerful message you can send to your listener is with the amazingly simple technique of repeatedly interrupting. This is discussed in greater detail elsewhere in the book, but here are a few principles to take note of right here and now.

When someone interrupts you, you know that they were not interested in what you had to say. When someone interrupts you, you know that they believe what they have to say is more important than what you have to say. When someone interrupts you, you know they think they are better than you!

When you are communicating with others, take a breath after your partner has finished their sentence and before you speak your next sentence. In that breath I know that you heard what I said, you are taking it in and appreciating my communication. This one thing alone is gold. To stop interrupting others could be the single most important skill you learn from this book.

The Complaining Communicator

Complainers are in the same trouble as the Baggage Communicators. You feel bad when you are around the complainers. When you complain, the state that you put your listener in is the state that they will connect to being around you. If you are a chronic complainer, you will instill constant negative feelings in others and will push people away rather than draw them near. Complaining is something best left for customer service problems and avoided in communication with those you love or would like to love!

The Gossiping Communicator

Gossip is probably the most deadly, miserable way to communicate. Don't use it, don't participate in it, and don't respond to it. You are giving away so much of who you are when you spread or even listen to gossip.

As tempting as it is, be bigger than that, starting today. If you are a gossiper, remember that there are others who have evolved beyond gossip. For those people who have risen above the need to gossip, they see you for what you really are. And here is my take on who you are as a gossiper. You are very insecure, your self-esteem is dependent on finding the fault in others, your world exists in a space of the small, weak, and petty. There is no bond that I would want to share with you by participating in your gossip. I know that anything I tell you will become public knowledge and be used against me.

Seriously evaluate any need you may have to gossip. Find out why it is important for you to talk about others in a way that is demeaning. Notice the reason that you need to spread bad news about others. If you are around someone who gossips, share your thoughts on gossip. When you state, "I really don't want to hear that, it is none of my business, and anyway I really like George," you encourage your listener to stop gossiping.

The key to great communication is mutual disclosure that is respected and appreciated. If you avoid the eight styles of negative communication and focus your attention on your partner, you will charismatically communicate with almost everyone you meet!

SEVEN

Positive Expression and Inner Magnetism

Be positively expressive in all situations and spark your inner magnetism so that everyone wants to be in your presence. Notice what it is about the people you enjoy being around that attracts you to them. It is how they make you feel about yourself. In the same manner, when you are upbeat and make the other person feel welcome and appreciated, they will be drawn to you – irresistibly.

Smile

One of the simplest and most effective ways to project your positive attitude and energy is to smile. A smile truly is a magical movement. It is so simple, so natural, and yet it creates a powerful chain reaction. First of all, when you smile, your body releases endorphins to your brain, which, in turn, make you feel better.

Just imagine the chain reaction that occurs when you smile. Your mood changes. That, in turn, affects the moods of those around you. That causes their actions and reactions to be altered in new and more positive ways. And, like a pebble that is tossed into a pond, the circle of influence just keeps on growing and expanding.

> **It shows you that you could make a difference in the world each and every day – every moment – just by putting a smile on your face!**

It's easy to see why a smile makes such a difference!

Being attracted to someone who is smiling is genetically hot wired into our unconscious minds. Studies show that babies, when shown photographs of faces, respond more readily and more positively to those with smiles than those that are serious or frowning. We are just naturally attracted to smiles.

A warm and natural smile indicates to us that we are accepted, welcome, liked, safe, and have approval. Of course that makes us feel good, and we are attracted to those feelings. So in becoming more attractive to others, it is easy to smile and share with them all those warm feelings.

Have you noticed that you can sense whether a person is smiling or not, even when you can't see them? When you are talking on the telephone, it is easy to become aware of this. When a person is smiling while talking there is a difference in how they pronounce their words, and in the inflection and tone in their voice. So we can see, feel and hear the effects of a smile!

Of course, if you are going to have a really charming smile, there are a few details that need to be remembered.

Clean and polished teeth are a must! Healthy teeth and gums indicate more than just how efficient your dentist is. They are also signs of the care you take in your personal hygiene and the likelihood of overall physical health.

It is a good idea to check your smile in a mirror after you eat anything – even after quick snacks. Food easily gets caught in teeth and we certainly don't want to be all dressed up, looking really sharp, and then have a piece of spinach stuck in our teeth when we flash someone our best smile!

Correcting crooked or chipped teeth, or filling in gaps between teeth, can be costly and time consuming. However, if it affects your overall attractiveness, it might be some of the best money you could spend. It is an investment in your self.

A pretty smile is also enhanced when your lips are smooth and moist. Chapped lips look sore and tend to crack when you smile. There are numerous products on the market to correct chapped lips. And drinking plenty of water will re-hydrate your body, as well.

Radiate Warmth

Imagine that you are the sun. Glowing, radiant, and brilliant. Just as the planets revolve around the sun, and people are drawn to stand near a fire, you will find that others will gravitate in your direction when you project radiant warmth.

Radiating warmth is a way that you can make others feel comfortable, welcome, needed and appreciated. It is the way that you draw a shy person into your conversation, the way that you listen to the words of another person, and understand their concerns. Everyone wants to be heard and understood.

The easiest way to give them that assurance is to make eye contact, nod your head occasionally, tilt your head slightly to one side, and make supportive comments. When you summarize their words and repeat them back to the person speaking, they know that you are listening. In this manner, you have the opportunity to be sure that you understand the message properly as well. They, in turn, can clarify any point that may be unclear or misunderstood.

We all enjoy being appreciated. Through appreciation we receive positive feedback that our actions, thoughts and intentions are recognized. It adds to our sense of self-worth in context to those around us. And as much as we like to be appreciated, so does everyone else.

Being appreciative can involve anything from a simple thank you to reciprocating with an elaborate gift. The token of appreciation should, naturally, fit the occasion. However, in everyday situations, it is usually sufficient to show appreciation in simple ways.

Again, a smile, a thank you, a nod of the head may be appropriate signs of appreciation. A compliment about something that person has done, recognition, or a handwritten note can all brighten another person's day. Being included in social plans, drawn into a conversation or entrusted with information or a project are all forms of showing appreciation as well.

> When you extend these types of warm feelings to others, they will naturally gravitate towards you, wanting to experience more and more of those good feelings.

The Golden Rule of Charm

Remember in Sunday school when you learned that the Golden Rule meant that you should do unto others as you would have them do unto you? The same rule applies in being attractive and charming. Always consider how you would like others to make you feel. Think about how you would like to be treated. What is it that would make you feel more comfortable in a certain situation?

When you answer those questions, you will know a great deal about how you can make yourself even more attractive than you already are!

Perhaps you can say something in a kinder fashion, or try to use a softer tone of voice. Maybe you could learn to verbalize the positive thoughts that you are thinking about people, but just never get around to saying. Notice something special about a person, or ask them about their family or a special interest that they have.

> *List the things that another person could do to make you feel more comfortable if you were at a social function where you didn't know*

158

> *anyone, or were feeling awkward and*
> *ill at ease.*
>
> 1.
>
> 2.
>
> 3.
>
> 4.
>
> 5.

Being reliable and responsible adds a great deal to a person's attractiveness and charm. It is an indication of the person's honor and integrity. It doesn't mean that you have to do whatever someone else says, or that you have to take responsibility for their issues or chores. However, when you agree to be somewhere, or take care of something, then it is important that people know they can rely on you keeping your word. Being on time for appointments, meeting deadlines, following through with plans that you have made will all leave a positive and solid impression on others.

Guidelines for Positive Expression

Avoid whining, complaining, criticism, bias and prejudices. When you speak in this way, it indicates that you are focusing on the negative. It becomes tiresome for the person listening to you, and they may get the impression that you will speak negatively about them also. These negative communications only detract from the person speaking. Their personality, energy, intentions and focus all become less attractive.

Speaking negatively even changes your facial features in an unflattering way. You will find that you frown, your jaw and facial muscles become tenser and your mouth curls in a sneer rather than in a smile. Whining, complaining, criticism, and talk of bias and prejudices tend to bring the mood down for both of you, leaving you both feeling worse rather than better. The other person will leave with an unconscious correlation that speaking with you makes them feel drained rather than energized.

> **When speaking with someone, think of how you can be additive to their life, leaving them with more than they had before the conversation began.**

Instead, ask people what they like about life, their activities, their favorite music or season. Focus on the positive aspects of their family, their work, and their hobbies. Think about what you would want others to ask you about. What would you enjoy sharing with someone in a light and upbeat conversation? Others might enjoy you asking them the same sorts of questions.

Can you think of something that would be immensely more interesting than the usual start up to a conversation? Something beyond, "Hello, how are you?" or "Isn't the weather nice?"

How about one of the following:

> *"The weather was so delightful today that I just wanted to take a long drive into the country and stop at one of those quaint little restaurants for a cup of coffee. What kinds of things would you like to do with a day like this?"*
>
> *"I was skiing last weekend and had an extraordinary experience. A white owl swept right over the slope ahead of me. It was beautiful. Isn't it wonderful when nature gives us such gifts? Have you ever had an experience like that?"*
>
> *"I notice that you are wearing a t-shirt from St. Thomas. What a wonderful place. What did you like best about your experience in the Caribbean?"*

Although each sentence involves you revealing something about yourself or your experiences, it comes around to bringing in the other person's experience or thoughts on a subject. While disclosing something about yourself, you are keeping the attention focused on the other person and showing your interest in them and in their experiences.

Generally speaking, when you strike up a conversation with someone, they will also be trying to interact, and therefore will be helping you out. When you find a way to get them to open up and talk about themselves, it will make it easier for them to respond and keep the ball rolling. You are offering to talk about a subject that they know best!

Speaking of which, it is a good idea to avoid talking to people about subjects that they have no knowledge about. When a person begins to speak of a subject that the other person knows nothing about, it can be like speaking a different language. Many areas of interest have their own lingo, such as sports, astrology, architecture, music, and on and on. If a person indicates that they know nothing of a certain subject, make a few comments and move on to another subject. Otherwise it starts to be a monologue or a lesson, rather than a conversation. That is, of course, unless the person has requested that you enlighten them on the subject.

It is important to be very aware and sensitive to the interest level of the other person.

While conversing, learn to look beyond the words that a person is saying. Become sensitive to what they really need from you. Perhaps it is sympathy or appreciation. Maybe they are seeking advice or a differing point of view. Do they need validation? Are they looking for a recommendation or solution to a problem? It could be that they just need to have someone listen to them.

When you develop a good rapport with another person in a conversation, you will become more and more aware of the underlying need or purpose of the dialogue, which often goes beyond the subject at hand.

Be fully present during a conversation. Have you ever had the experience of speaking with someone and their eyes were wandering off in different directions, or they didn't appear to be listening? Perhaps they were keeping themselves busy with other activities while you were speaking. It doesn't feel very good, does it?

In this busy world, we often have to multi-task, especially if we are at work. However, in a social situation, it is only respectful to give a person your full attention in that moment.

If you are not interested in speaking with the person, or the subject matter doesn't interest you, find a gentle yet effective way to end the conversation or change the subject. If your time is limited and you can't give the conversation your attention, ask the other party if they would agree to reschedule the talk for another time.

To be fully present, you should stop what you are doing, turn your attention to that person, have eye contact with them and have mental contact. Acknowledge their words and give them feedback.

Because you are a charismatic person, you would never dominate the conversation. Remember that a conversation is not a monologue or dissertation. Lecturing a person will remind them of a longwinded professor at school, or link them to memories of getting a lecture from their parents. Neither is very desirable when making light conversation in a social setting.

Think of a conversation as a ping-pong game. Take your turn and send the conversation back to their side of the table. Repeating that, back and forth, throughout the length of the discussion. Taking turns allows both parties to participate and will keep you both interested and engaged. Be as interested in their words and thoughts as you would want them to be in yours.

Respond throughout your conversation with comments, a head tilt, or a nod. Hopefully if your eyes gloss over and your head starts to bob, they will realize that they have exceeded your attention span. And naturally, if you notice the same in your conversation partner, you will know that it is past time to turn the conversation back to their side of the table.

Avoid offering advice. As tempting as it is, there is little good that comes from freely giving out advice. First of all, when a person is telling you something, they generally just want to pass on the information, or get something off their chest. They are looking for an ear to hear them, a sounding board. Giving them advice undercuts their sense of confidence.

When you offer advice, you are taking responsibility for the outcome of the issue. If they take your advice and things don't turn out well, you have become involved in a negative way. Remember that rarely do we know as much about a situation as the person who is directly involved in it. All of our own advice would be prejudiced to our own experience, and, at best, poorly informed.

How do you feel when you are telling someone a story about an experience and they start dishing out advice to you? Oftentimes, their words seem redundant and possibly belittling.

If someone actually asks you for advice, instead of telling him or her what to do, you might tell him or her what you have done that has worked for you. From there, they can determine what it is that they want to do for themselves. That alleviates any responsibility that you might have had, and honors their intelligence for being able to figure out the solutions to their own problems or issues.

When it is necessary to deliver a bit of criticism to someone, it is kinder and much more easily accepted by that person if it is sandwiched in between two compliments.

Notice the differences in feeling when delivering the same message by two different methods. Which do you think you would more readily listen to?

164

> *"These files are a mess!"* **Or** *"You have probably already thought about this, but the files will have to be organized in a more practical way. You always do such a good job with these things. I know that you can improve their present condition."*
>
> *"I don't like your hairstyle."* **Or** *"You have such lovely hair and your eyes are so pretty. I wonder if your eyes would be enhanced if you were to wear your hair in another style."*
>
> *"You need to lose weight."* **Or** *"I've noticed that I have started to show a couple of extra pounds. I wonder if I go on a diet, and begin an exercise program whether you would be interested in doing it with me. You have more discipline than I do, and you would be a great inspiration for me."*

Sometimes it seems awkward to speak like that, but after you practice it, the wording becomes more natural. Another person is not likely to notice the roundabout approach to the suggestion, as much as they would notice how badly they felt had you said something more critical and blunt.

If there is something that you have wanted to point out to another person, but have been trying to figure out a kind way of addressing the issue, try this technique:

Write down exactly what you would like to say. Be blunt and to the point. Say whatever comes to mind.

Write it out again – even a couple more times. Just get it out of your system first. Sometimes we build up so much energy around a subject that when we finally blurt it out, it comes out much worse than we wanted it to.

Next, put a smile on your face. (You remember what that does for you, don't you?) Now, while smiling, think of another way to write down the same idea, only in a kinder, more gentle fashion. How is that beginning to sound now?

Now, think of a person who you highly respect. This person could be someone you personally know, or some famous person. You might think of your favorite grandparent, Jesus, Buddha, a teacher, or anyone else that might be important to you.

If you were to make this comment to them, or in front of them, how would you word your criticism?

In what way do you see your comments changing now?

You may now have found the perfect way to deliver your message, or you may have to continue wording it and seeking improvement. Once you have a final draft, you could still present it to another friend or colleague to get their opinion of what you have written. Oftentimes, how we think we are saying something is not exactly how another person will receive it when they hear it.

The above technique is also highly useful when communicating with the printed word. With the advent of the computer and especially the Internet, we are communicating in writing more rapidly than ever. It may be one of the most valuable tools that we have for transmitting information. However, there are distinct pitfalls.

When we write, we cannot transmit our tone of voice and our inflections. Words meant to be said with a smile, or a chuckle, may fall flat on the printed page. Oftentimes meanings are mistaken and feelings are hurt. So once again, in this case, re-read your words and think about how your words might be "heard" if that other person is in a different mood or frame of mind than you are.

Don't Steal Another Person's Thunder

When in a conversation, keep the spotlight on that other person. It doesn't mean that you can't talk or tell a story, or interject with comments about your experience. However, when a person is telling you a story, refrain from trying to outdo them. We are reminded of "fish stories" – each person's fish being larger than the previous person's.

This premise goes back to appreciating the other person, giving them the feeling of value and worth. You will be viewed as most attractive and charming when you give that person your full attention, and join in their enthusiasm regarding their experience.

An example of how you could do that is as follows. Say a person is telling you about their recent vacation to Fiji. You could respond with, "Yes, I've been to Fiji also. What was it YOU liked best about it?"

Showing that you care about another person's interests is very much appreciated by the people that you interact with. Take a moment to ask about their spouse and their children (or grandchildren). Ask how their health is, or how they enjoyed their latest skiing trip. Ask if they have any new additions to their favorite collection.

In this way you show them that you have paid attention in the past to the things that they have said, and the interests that they have. It shows that you care and are concerned about them.

Projecting Positive Energy

Do you know how some people leave you feeling drained while others leave you feeling pumped up and energized? We, as attractive and charming people, always want to leave another person with energy and a good feeling about us.

Sometimes we are tired or drained ourselves and it is hard to muster up the excitement or energy to project and share with others. However, there are a couple of simple methods that can ensure a more positive result in your conversations.

When you are talking with a person, find three things that you like about them. It can be anything. Their clothes, their eyes, the way they express themselves. It could be something they have accomplished, or the way they talk about their future dreams. Whatever it is, find three things. No matter how obnoxious some people can get, you can always find three redeeming qualities.

When you are concentrating on their good points instead of on the reasons why they irritate you, your own energy actually changes. You become more positive and the energy that circulates between you is certainly more positive. Your body language will change, the expression on your face will change, and who knows, maybe your attitude about this person will change! The only way to demonstrate this principle is to try it out.

As you find these good points in another person, try to incorporate them into your conversation. This creates an even greater sense of positive energy and you will, personally, demonstrate great charm, becoming more and more magnetic to all around you.

Try out some of the following phrases.

"That is a beautiful necklace. Did someone special give that to you?"

"What have you done lately that is fun, to keep you looking so happy and relaxed?"

"I see you have a beautiful wedding ring – do you have children?"

"I like your ball cap. Have you been to a game lately?"

"I liked your ideas on that report. Do you have any other suggestions?"

You get the idea. It is very simple, but so often overlooked. Try incorporating compliments into your conversations and notice the wonderfully positive reactions that you get from the people you are speaking with.

Another way to raise the energy levels between you and your conversation partner is to imagine an electric current running between the two of you. Whether it is when you are shaking hands or when your eyes are in contact, feel that electricity. Visualize it and project it.

The Voice

One of the fundamentals of conversation is the voice that you use to communicate with. It is your voice that carries the words.

You can probably remember when you were a child and your mother would call your name. It was your mother and it was your name. But the tone of voice told you everything. You knew if you were in trouble, or being called to dinner. You could tell if she was playing with you or giving you orders. It all rested on the tone of voice.

> **We can consciously alter the way we sound when we speak. Have fun practicing different ways of using your voice.**

There are so many ways to use the voice, that it would be a book in itself to try and cover them all.

However, there are a few details that we can briefly review here, that will help you to put your most charming voice forward.

Volume

Sometimes we have to shout to get attention, or be heard over other noise or across a distance. Other times, when you are positioned close to another person, it is a good idea to gauge how loudly you really have to talk. If you are in a quieter restaurant, you may not want everyone to be able to hear your conversation. And it is most probable that they don't particularly want to hear it either. When a person is close at hand, it may actually be possible to whisper to them, and they will hear you. Whispering is not generally necessary, but certainly if a whisper is sufficient, why would someone be talking loud?

We forget that we don't have to talk loud. If we have been listening to a loud car stereo, or traffic noise, or have been working around loud machinery, we simply get used to having to speak up – or it could be that our hearing is not registering at normal levels.

Whispering can be very alluring. It will draw a person closer to you – while speaking loudly can be very abrasive and repelling. Try whispering and see what happens around you. A soft-spoken person is just naturally more attractive, to both women and men.

It is interesting that when you get a cold and have to whisper or talk softly, everyone around you starts whispering back. It is an interesting experiment in that you begin to realize how softly you really can speak and still be heard. And it is fascinating how people will automatically take your lead and speak much softer back to you.

Everyone can alter his or her voice, to a degree. You can make it sound deeper, more resonant. You can change the qualities of the tone. Being a hypnotherapist, I (ML) notice that I have my regular voice and my "hypnotherapy voice". The voice that I use when I am in a session with a client, or when recording my audios, is much softer and mellower. We can consciously alter the way we sound when we speak.

Voice Tone

How do you sound? Do you sound to others the way that you think that you sound? Probably not. Have you ever heard your voice on a recording? I have rarely heard anyone say "Oh, that sounds just like me." Most people can't believe the way they sound. It isn't anything at all like they thought that they would sound.

It is worth the effort to work on improving your voice tone. Speak into a recorder and listen to it over and over. Make different recordings, slightly altering your voice tone until you find one that you like. Then practice using that voice on a regular basis. It will seem "fake" or forced to begin with, but over time you will, at the very least, be able to turn it on when you want to.

Your personal attractiveness will rise when you go out of your way to be helpful and courteous to another person. You also have heightened appeal when you show respect for the environment, animals and life itself.

Showing small courtesies takes a minimal amount of effort and yet the rewards are high, both in how you feel about yourself and how others view you.

Pausing to hold a door open for someone, stopping to help someone who is fumbling with a heavy load, or slowing your steps to accommodate a slower moving person all show respect, care and courtesy to others. It demonstrates that you are not only focused on your own life and activities, but also taking into consideration the experience and needs of those around you.

Self-confidence and Self-esteem

Why do we find a person with high self-esteem to be attractive? What is it about them that draws our attention and our admiration? Is it the mystique? Is it an aura?

A person who exhibits strong self-esteem is telling the world that they value themselves. After all, the meaning of "self-esteem" is the esteem (value) of self. It is the estimation of worth that you are giving to yourself. So when a person recognizes their own self-worth, and exhibits that to the rest of us, we start to think that they know something that we don't know. In other words, they think that they are special, and have value.

Likewise, when someone shows the world that they have low self-esteem, we figure that if they don't think so highly of themselves, then why should we be impressed or respect them?

In both cases, we (the observer) simply go along with the estimation that the person has signaled us is valid. We tend to just believe the verdict that the person has put upon himself or herself.

So why is that attractive? We, as humans, are naturally attracted to that which has been deemed valuable. We also tend to want to be a part of a larger group. We often follow the lead, join groups, and go along with the majority opinion – just to be part of the group.

High self-esteem can also create an illusion of attractiveness, or competence, even when it perhaps isn't there. It is possible for us to be fooled. After all, attractiveness is a subjective attribute.

Take the case of Miss Piggy of the Muppets. She was confident and had great self-esteem. She thought she was beautiful and she was extremely dramatic in the display of her illusion. Everyone fell for it. And Kermit the Frog even fell in love with Miss Piggy. But the truth was – she was a pig! She was not the curvaceous Jessica Rabbit, of "Who Framed Roger Rabbit" fame. She was the roly-poly Miss Piggy.

So developing self-esteem and confidence can distinctly increase your ability to be irresistibly attractive. It can create an illusion or aura of value, worth and desirability.

Self-esteem, as we have said, is the way that you feel about yourself. Self-confidence, by contrast, is the way that you feel about your abilities. We will talk about self-confidence in just a moment.

To develop your self-esteem you need to be doing things that make you feel good about yourself. No amount of compliments will give you self-esteem. You can't have enough money, or buy enough things, to build self-esteem. You have to feel it from the inside. Your self-esteem will be directly tied to your values and how you act upon them. It requires that you live a purposeful life. That, in itself, will require that you have a dream, the courage and conviction to go after that dream, the discipline to stay true to it, and the stamina to see it to fruition. It can be a monumental dream, or it can be simple. It doesn't matter. It simply has to be your dream.

To have that dream, and to go through everything it will take to get to the fruits, will require that you stay true to yourself, keeping your sites on your life and your goals. That process will build character, sharpening facets of your personality and honing you into the best person you can possibly be.

And when you have done that, or while you are still in the process of that, you will naturally be rewarded with a strong sense of yourself. You will be proud of who you are. And what is that if it is not self-esteem?

Self-confidence

Self-confidence is built through practice of a skill. We may have confidence in shooting hoops on the basketball court, or self-assurance in asking a person out on a date. We may feel confident of making a sale at work, or sure of ourselves giving a speech.

We can build self-confidence through visualization, repetition, discipline, study, and practice. The more we do something, the easier it gets, and the more skill and precision we develop.

Having developed our skills in one area, our self-confidence may flow over into other areas. In other words, our skills in driving a car may lead us to be confident in our capabilities to learn to fly an airplane.

> *The more areas in life in which we have developed confidence and competence, the more likely we are to be assured in our abilities to take on new and greater challenges.*

Because this confidence generalizes to other areas of our lives, having built skills and competence in any area, which in turn increases our sense of confidence, allows us to face difficult challenges in our lives. In turn we are more likely to take better care of ourselves, standing up for what we believe in or going after what it is we might need. We are better able to create personal boundaries and defend them.

You can easily see how this, by nature, allows us to maintain our values, and create and live out our dreams, providing us with the purposeful life that, ultimately, gives rise to our positive self-esteem.

We come full circle. Self-esteem and self-confidence continually feeding each other, supporting each other, spiraling upward, stronger and stronger.

So you can see the importance of building skills and engaging in activities that will give you positive experience and practice. The more we shy away from trying something, the more we hide from life and its experiences, the more hesitant we become. We lose confidence, begin to have self-doubts, and engage in negative self-talk. We become afraid.

In our fear and self-doubt, our shoulders begin to sag, our head drops down a bit, and we may shuffle when we walk. We may appear to be weak, and we become prey to bullies, aggressors and abusive people.

Granted, everyone has experienced self-doubt, nervousness, and uneasiness to some degree. In fact, we all have a range of reactions to different circumstances in which we find ourselves. When we are in the middle of it, we may think that we are the only person in the world who is feeling this – that everyone else is more confident and in control than we are. However, aren't we surprised when we find out that everyone has and does experience those moments of insecurity?

Some people experience this nervousness and self-doubt on occasion, in certain circumstances. And there are others who simply lack any self-confidence or self-esteem at all. People who truly lack self-confidence will exhibit signs of compensation that will include:

- *Being boisterous – they need to boast about things to impress people.*
- *Grabbing attention – they try hard to keep themselves at center stage.*
- *Always having to be right – admitting they are wrong would show a weakness.*
- *Belittling others – making others appear in a negative light gives them the mistaken perception of being better.*
- *Exaggerating – they try to make things seem bigger or better than they are.*
- *Bullying – they compensate by being aggressive and making others fearful.*
- *Being nervous and tentative – they look to others for approval and are afraid of making mistakes or losing the respect of others.*

There are other signs of low self-esteem and self-confidence that are exhibited through a person's body language. These are discussed in other areas throughout the book.

Low self-esteem and self-confidence are fear-based. Our fear begins as a small seed, but, if left to fester, can grow out of proportion. It can overwhelm us and paralyze us. The more it is allowed to grow, the more difficult it is to combat it. There is only one true way to conquer your fear. Face it and move through it.

The antidote for fear is love. And that love embodies trust, security and confidence. It is important to find and build these qualities inside of you. To do this you may practice one or all of the following (the more, the better):

- *Reward yourself along the way – and make it fun.*
- *Put yourself in situations that are affirming and remove yourself from ones that are belittling, humiliating or in some way drag you down.*
- *Be excited about your successes, your self and your life.*
- *Accept people's changing attitudes and reactions to you.*
- *Be prepared to change friends, lifestyles, etc., if needed, to step out of the old frame of thought.*
- *Constantly give yourself new challenges as the old ones are completed. The more we stay in our comfort zone, the less confident we are about stepping out of it.*

Failure and rejection are very difficult experiences to face and live through. However, it is really a numbers game. When you save yourself from the experience and only pop your head out once in a while to take the risk; if your attempt fails, you are "one for one". It is a 100% failure rate. That looks very bleak and we are certain that we have just proven to ourselves, and perhaps to the world, that our fear of failure and rejection is well founded.

However, if people were to take risks all the time, just putting themselves out there and going after what they want, the chances of success increase dramatically. If you take one hundred risks during a week and have ninety-nine failures, that is a 1% success rate, as opposed to 0% in the previous example. If ten of those attempts turn out well, you have just gotten yourself a 10% success rate!

Beyond this hit and miss type of risk-taking, what if you actually began to learn something from all the mistakes and failures? What if, with each rejection, you learned to tailor your technique just a little bit toward those factors that have resulted in success? Pretty soon, the success rate is increasing faster and faster. After a while you may figure out exactly what works and what doesn't. Then you can simply use those parameters that have given you the results that you are looking for.

In my practice I have worked with several clients who have been afraid to approach a member of the opposite sex for the purpose of asking them for a date. These represent cases of classic fear of rejection.

There is really no way around this issue except to go headlong into it and take advantage of the increased odds of the "numbers game". You may just plunge in and actually do it. This would mean that, for someone in our dating example, they would start asking out as many people as they could. Another way to begin the transition into someone who is comfortable with his or her former fear of rejection or failure, would be to utilize a powerful visualization technique referred to as desensitization.

Desensitization Exercise:

One of the most powerful ways to overcome fear is through a visualization technique for desensitization. Short of actually going through a dreaded experience, one can visualize going through it until it becomes less threatening or overwhelming. Then, when they experience the event in reality, it becomes much easier. It is quite simple and can be individualized to any specific goals.

You may read it over and do it on your own, or you may want to record this script, slowly, onto a tape and listen to it.

> *Close your eyes and take a deep breath. Hold the breath in for a few seconds. Exhale and allow your body to relax. Just let go of any tension that is stored in the body. Feel your body as it relaxes. Take in another deep breath, holding it. And once again, release it, and relax even deeper.*
>
> *Begin to imagine yourself at the beginning of the experience that is giving you such fear.*

Begin to move through the scene, noticing any details of the episode, until the event is completed. Notice how you feel when you have completed the entire experience.

What have you learned by going through it? What would you do differently?

Once again, imagine going through the same experience, playing it out. You can make any changes to the scene as you move through it. Notice how it feels this time. What has changed in the event as you move through it a second time? How do you feel when you have moved completely through it?

Once again, start at the beginning of the episode. Move through the scene again, noticing the details of the experience this time. How do you handle it differently? How do you feel as you move through it this time? Has the outcome changed at all?

Continue to repeat this procedure until you can look at this anticipated event and find that you are more comfortable with it.

Some events will simply never become completely comfortable. However, through this desensitization technique, you can gain some familiarity with the process of moving through it, and get different perspectives of what may happen.

> *You have the opportunity to try out different reactions to the events, handling communication and action in various ways, perhaps creating different outcomes.*

This technique is most often done as a visualization or meditation. However, if your mind tends to wander and you are having difficulty concentrating or seeing and feeling the details, you may want to try doing this as a writing assignment.

Simply write a story about this fear, including details of the story as you move through the scenes. Connect as fully as possible with each emotion that you would experience, as though you were actually there doing it. Continue to write the stories, over and over, until the emotions around the fear have subsided. Again, notice how the story changes each time it is told.

Overcoming these simple, yet paralyzing, fears will change the way that you live your life. You will make choices more aligned with your values, you will increase your chances of obtaining what you desire, and you will increase your self-esteem and self-confidence – and hence, become more irresistibly attractive!

Attracting Your Life Partner

What does your life partner have to do with self-esteem and self-confidence? Everything!

The value that we place on our self is the type of person that we think that we will be able to attract. If we feel that we are not worthy, we will settle for less, or set our sights on a person who is not right for us.

Furthermore, when we are living out our purpose and full of self-confidence, we will be circulating among people with similar values and interests. We will be displaying our true selves. This will make it easier for someone who is attracted to us, for the person we truly are, to actually find us. If we are hiding behind a mask, pretending to be someone who we think society will approve of, or that represents a script that our parents created for us, our true loves will not be able to recognize us. We probably won't even be in the same social circles.

So by shining our own brilliant light, we have the optimal chances of attracting the perfect mate into our lives.

Let's find out whether you are shining your light and who your ideal life partner might be. Answer the following questions:

> *When you were a child, what was your favorite game?*
>
> *When you played "make believe" who was your favorite character to play?*
>
> *When you were a child, what was your dream for your future?*
>
> *What is your dream now?*
>
> *What would your dream be if there were absolutely no restrictions on what it could be?*
>
> *What activity do you do, or wish you could do, that really excites you?*

Who would you like to spend the day with if you could chose anyone, living or dead?

Why?

What is it about them that you admire?

If you could be just like them for a day, what in your life would you change?

Why?

In what way could you change it now?

When you are in your final days of life, what will you have wished you could have or would have done?

What stops you from doing it now?

When you die, what would you like people to say that you accomplished?

When you die, how would you like your epitaph to read?

Imagine entering a room where all of life's necessities were taken care of. There would be no need for a job to sustain you and you could have everything it would take to keep you entertained and fulfilled. What would you find there?

What have you learned about your life purpose, or at least about your personal dreams and goals?

In understanding more fully where your passions and goals lie, you can move towards fulfilling them. As you do so, you will be more apt to attract your ideal life partner to you. Remember that you have to be careful about what you wish for – you just might get it! So it is wise to have a clear idea of the traits and characteristics of the person you want to share your life with.

To assist you in making that determination, please answer the following questions:

How would you describe your ideal mate?

What characteristics would you want them to have?

In what ways will they be similar to you?

In what ways will they be different from you?

What is it in you that this person will find attractive?

What is it in you that might prevent this person from finding you attractive?

What can you do to more fully be attractive to your ideal mate?

In what ways will your life be better when you are together?

In what ways will your life be worse when you are together?

In what ways do you feel yourself blocking them from coming into your life?

In what ways do you feel outside forces blocking the way for you to be together?

Is there anyone in your life who meets the criteria for being your ideal mate?

How will you know that you have met your soul mate?

What would you be willing to do to be better prepared to have your ideal partner in your life?

When we meet our life partner, or someone who we would really like to spend time with, it is vitally important that we continue to practice our attractiveness techniques. Once we have brought them into our lives, and both parties have chosen to make a commitment, we want that other person to choose to be with us each and every day.

Setting standards and abiding by them certainly makes you more attractive. When a person has no standards, it is no compliment to be their partner or lover. It is when you have set the highest of criterion for your partner, and that other person meets that value, that it becomes an honor for them to be with you. You honor yourself in setting those standards, and you honor your partner when you accept them into your life. And the reverse is true.

When you find someone who has set high standards for themselves, it will be more challenging, and rewarding, to meet that level. It is when we far exceed those standards that we begin to feel that we are in a relationship that is "beneath" us. That phrase simply means that the other person has values that are less stringent than the ones that we place upon ourselves.

So, among all the other criteria that we seek out in another person, an important one is finding similarities in value systems. That is a large part of the spiritual aspect of finding your life mate.

In the above questionnaire, there were many questions relating to what you are hoping to accomplish in this lifetime and whom you are seeking for a partner. Keeping in mind all the answers that you gave for those questions, take a few moments to do the following exercise:

Stepping into You

Find a comfortable place to sit, free of distractions. Close your eyes. Take a deep breath and hold it for a few moments. As you exhale, allow your body to fully relax, letting go of any tension. That's right, just relax.

Now, take in another breath and hold it. Hold it. Now exhale, and again relax, this time even deeper. Only you will be able to allow yourself to fully relax in the best way you know how.

Begin to imagine yourself outdoors. It is a beautiful day and the landscape around you is just the way that you would imagine it to be. You are comfortable here and it is relaxing just to enjoy this place.

As you look around you notice that there is a building nearby. You are drawn to that building and find it to be a very curious place.

That curiosity urges you to open the door and enter the building. As you do so you find yourself in a room. That room is exactly the way that you would like it to be. You spend a few moments exploring the room, discovering all sorts of wonderful items and furnishings.

As you continue to look around this room, you discover that there is another "you" in the room. A "you" that is exactly the way that you would like to be. The new you with the highest possible standards, and with the behaviors that you admire and desire. The you that is fulfilling their life goals and purpose.

You may be a bit surprised to find the other "you" here, but soon you find it simply fascinating. You begin to observe the other "you", noticing how "you" walk and move. You notice "your" energy and behavior, admiring "your" positive qualities.

That other "you" walks to the center of the room and turns their back to you. You walk up behind them and stop, standing just two paces behind them.

Taking a deep breath and relaxing, you step forward and meld right into them. You become one with them, feeling your hands in their hands, your feet in their feet, and looking out through their eyes.

Take a few moments to feel what it is like to be in their body, this new you. What do you notice about the way you feel? What do you notice about the way that you are thinking? How is it different to be in this aspect of you, rather than your previous you?

Take a few moments to move around, explore your surroundings and truly experience this new aspect of you for a while. Take as long as you would like.

When you have finished exploring, notice what you have learned about yourself. What is different about yourself now? What is the same?

In what ways would you handle your life differently if you were in this new aspect all the time? What would change in your life if that occurred? What outcomes would be different? How would people react to you differently? How might your goals be more easily attained?

In a few moments you will have the opportunity to step back out of this new you. You always have the option of just staying in this new aspect of you. You don't have to step back out. And you can also choose to step back out and bring with you certain qualities of this new aspect of you. Regardless, you will always be somewhat affected by having experienced this visualization.

So, deciding whether to step out of this aspect, or not, or just bring back certain qualities, simply count 3-2-1, and make it happen. Very good. How are you feeling?

Now, remembering all that you have experienced in this exercise, simply count from one to five, emerging from your visualization and returning to full wakefulness in this present moment. One, two, three, coming up, four, aware of this time and space, and five, wide aware and refreshed and returned to this present moment.

Oftentimes we are too eager to remember our shortcomings and dismiss our successes in life. In our feelings of being wounded, we may count our shortcomings rather than our blessings. However, that is simply a matter of perspective. It is what we are focused on. And it has been proven that what we are focused on is what we tend to make of our lives. A negative attitude will drive us deeper into poor self-esteem and lack of confidence, creating a world of incompetence and failure. While a positive attitude can support our increasing self-esteem and confidence, furthering our competence and success. It is a choice. Which sounds more appealing to you?

Is the glass half empty or half full?

For the purposes of raising self-esteem and confidence, you may want to make a recording of the following guided meditation for attaining an empowerment symbol.

Attaining Your Empowerment Symbol

This meditation will give you a metaphorical tool that will anchor a positive feeling of self-confidence. By anchoring, we mean that by connecting an object, scent, or touch with a feeling, a person can revisit that feeling whenever they come in contact with or think of that anchored object, scent, or touch. In this visualization, we will anchor a symbol that you can think about that will allow you to regain positive feelings from the past.

You may begin by taking in a deep breath and relaxing in a comfortable location, free of distractions. Take in another deep breath and hold it for a moment. Hold it...now, as you exhale, allow all the tension in your body to just flow out of you. Relaxing even deeper. That's right.

Allow your mind to drift back to an earlier time when you really felt good about yourself. A time when you felt in control of your life and that you were making good decisions. It might be an event in school, at work, or in your personal life. Fully remember that event and fully connect with the feelings that you had about yourself at that time. Notice how your body feels. How does your energy feel? So proud, so good. That's right, really feel it.

Continuing to feel these good sensations, allow your unconscious mind to create for you a symbol that you could hold in your hand, that would represent this good feeling. Just allow it to come naturally. It

192

*could be a sun, or a star, or whatever
comes to mind.*

*Examine the symbol that you have
in your hand.*
What color is it?
What shape and size does it have?
*What does the weight feel like in
your hand?*
*What else do you notice about this
symbol?*

*While holding this symbol, think of
a time in the near future that you
anticipate having to go through
something that you are uncomfortable
with. It can be any situation where
you feel a sense of self-doubt or
lacking in confidence. Think of that
event now, while holding your symbol
of empowerment. Allow yourself to
move through the anticipated event.
What do you notice about it now that
you go through it with your symbol?
How do you feel about this event now?
What seems to have changed?*

*Once again, holding your symbol
in your hand, start at the beginning of
that event and move through it one
more time. How does it feel now to be
envisioning that event? In what ways
has it changed when you view it this
time? How do you feel about yourself
as you deal with the situation? How is*

*that different from the way that you
felt about this before?*

*If you choose, you may go through
this scene one more time. Notice any
further changes that have occurred in
the event or in your feelings towards
it. Very good.*

*Knowing that you may always
access this empowerment symbol,
anytime in the future when you need
to remember these good feelings of
self-confidence and self-esteem, you
may count from one to five, and return
here to the present moment, refreshed
and relaxed. One, two, three, coming
up, four, aware of this time and space,
and five, wide aware and returned to
the present moment.*

This anchoring tool, the empowerment symbol, is
as handy as it is powerful. You don't need any
materials in order to tap into its power. You simply
have to remember that you already hold it in your
hand. Have fun with it - the more you use it, the more
potent it becomes!

The Attraction Cycle

Remember that attraction is an etheric magnetic
quality that a person has. You have every bit as much
potential for being irresistibly attractive as the next
person. It is a matter of assessing who you are. When it
comes to physical beauty, we can make grand changes,
yet even those are within a certain natural boundary.
However, with self-esteem and self-confidence, there

are no limits. Whatever you can imagine, can be attained.

Set your sights high, raise your standards, maintain them and never compromise. Be proud of who you are and if you are not, then change whatever it takes to make it so.

Challenge yourself to take calculated risks and go after life - this is the only one you have this time around. Whatever opinions that others have, or any other imaginary block preventing your growth and progress, certainly cannot be more important to you than your own desires, goals and life. If you allow those things to get in your way, then it is exactly that - you have allowed it. It is your choice. Therefore, you may choose to be the one that chooses what happens in your life, or you may choose to be the one that lets life happen to them. Maybe you will be the one to explore the frontiers of the possibilities that this life has to offer you.

> *Imagine how you would manage your life if it were a game of chess.*

Dating

For some people dating is natural and fun. For others it is one of the most difficult activities that they can engage in. It can be complex and it can be perplexing. Yet, with some guidelines and the desire to circulate and meet people, dating can turn into a worthwhile and rewarding experience. How else can we discover that perfect life mate?

From the first glance to the first date, from the building of a relationship to making a lasting commitment, the stages of dating can be fascinating, exciting and energizing. And although we all have our eyes on the ultimate goal of finding and securing that perfect union with our most compatible life partner, we have to stop and smell the roses, and realize that getting there really is half of the fun.

In our infatuation, or our need to be in a relationship, some steps in the art of dating are frequently skipped. Some people will meet an attractive person and want to "work on the relationship" right from the start, before taking the time to determine all their compatibility issues. Other people, through fear or reluctance to get further involved, never get past the first stages. They would prefer to just keep sampling the menu and not "get stuck" with the same item over and over again.

It is a matter of personal taste and style. However, following the natural "dating curve" can be rewarding and a lot of fun. And throughout the journey we have opportunities to develop our charisma, maintain our attractiveness and let others discover how very irresistible we are!

Before we get into dating skills, let's take a closer look at what you are looking for.

Secrets of Finding the Love and Experiencing the Intimacy You Deserve

Communication with a special person or persons is going to play a large role in the quality of your human existence. It's possible that the most wonderful experience you can have on this planet is the involvement in a romantic relationship.

When you fall in love with someone, there is a euphoria that is experienced within that is rarely ever matched in terms of quality of experience. When you "fall in love" with someone, you are also often blinded to many of that person's faults or negative behaviors. This can be good or bad depending upon how the relationship emerges and grows.

As the relationship does grow, the problems your lover has kept silent about eventually surface, and you become aware that there will need to be "work" in the relationship. What makes the love bond grow deeper is that upon the revelation of the faults and problems of the other person, we still love that person and appreciate their ongoing love of us. Lasting love can be euphoric but it is really identified by the theme of acceptance and lack of criticism.

In this chapter let's start at the "beginning". If you are currently unmarried or are unspoken for, and wish to be married or spoken for now you will see just how to accomplish that desire. You are about to discover the very simple way that successful relationships begin. If you are currently married or spoken for, this chapter will help you turn the difficulties of the current relationship into the seeds of a happy and fulfilling experience.

Who Are You Looking For?

People who are in a "good" relationship knew at one time just what they were "looking for" in a relationship. It may have been conscious or unconscious. The criteria may have been physical, emotional, mental or spiritual (or some combination of these four) but they did know. They knew what kind of person they wanted to attract into their lives. In some cases they knew exactly which person they wanted to be with!

Many people, when beginning their search for a love partner, look "out there" and hope to find someone and they wonder why the person of their dreams isn't there. One problem is now in place that cannot be immediately overcome. When someone is looking for *any* another person it has the same effect as looking for no one. No sense of attraction is emanating. When people begin "looking for" specific traits or features, whether physical, emotional, mental or spiritual, there is a magnetic-like filtering process that attracts the people their way. By analogy, if you go driving in rush hour just hoping to see the "right automobile" for your next purchase, you will see nothing. If, on the other hand you decide to look for a specific car, you will see dozens of them on your way to work. This is similar to the experience we have all had of purchasing a car and then seeing dozens of them on the road. What happens is that a part of our brain now recognizes the new vehicle as significant to the self. You identify with it.

The very same mental process helps us recognize the people we want to invest time and energy into in a relationship. Unfortunately many people don't isolate the key traits of those individuals. If one determines the characteristics of the people they want to build relationships with, their internal processes will screen out the people who don't meet these criteria and will filter in the appropriate potential relationship partners.

If you are in the "looking" stage (and we are all looking for some kinds of relationships most of the time), then just what kind of person(s) are you looking for? We all need to engage in relationships that will help us grow as people. This means we need to be with people who have some similarities to us and some significant differences. People need to complement each other as well.

Some people are good with money. Others are not. Two people who are not good with money in one relationship could be a recipe for failure if a solution isn't found to take care of this difficulty. Many people can't stand the sound of a crying child. Others find it to be just another noise that kids make. If two people marry who can't stand the sound of crying children, what happens?

How can we attract the right people into our space? How can we spot the people we really want and need from the masses?

We must have some level of knowing what needs we want met and what we have to offer others. Many people haven't thought about the characteristics that they are looking for in other people for personal or even business relationships.

Those who tend to be looking for certain traits, behaviors and characteristics tend to find what they are looking for. You can thank the miraculous construction and functioning of your brain for this!

Exercise:

If you are in the "looking" stage for a partner, write down the key physical, emotional, mental and spiritual traits you will be looking for in that person.

> *Be certain to include everything that you think of. If your partner needs to make a lot of money or you don't want to pursue the relationship, write it down. There are no wrong characteristics to include.*

Having seriously evaluated what you are looking for in a special person on all levels allows your brain to begin filtering out people who don't match your criteria. This is a very important function for your brain to have because it makes the task of discovering possible life partners very simple for the brain. It knows who you are looking for and will now do the work for you.

In all relationships, there is a sense of give and take. There is a certain unwritten scoreboard of contributions to the relationship that are unconsciously evaluated by each person. When both people contribute fairly equally to the relationship, the relationship is generally happy and content. When one person contributes far more to the relationship than the other, the relationship will probably end or be very unhappy.

You have recorded those characteristics and traits that you would like to see in a potential partner. Now it is time to put the spotlight on you.

Exercise:

What do you bring to the relationship for the other person?

> *What benefits, experiences and resources will they gain from you?*
>
> *What do you see as your physical, mental, emotional and spiritual strengths? Write down all that you will invest into this soon-to-come relationship.*
>
> *Now comes the hard part. What inherent problems do you bring into a relationship with a life partner? (Are you in debt? Are you outside of this person's religious faith? Are you obese? Are you unhealthy? Do you have an unstable career or no savings?) Write down all of the areas in which you are deficient and be as thorough and honest with yourself as you possibly can be.*
>
> *For each item that you listed as a deficiency now, once again note the item in writing and then explain to yourself what you plan on doing to correct this deficiency. If it is impossible to change, simply write that fact next to the item.*

Having now taken an honest and penetrating look at yourself, once again turn your attention to your potential life partner descriptors in the first exercise. Some of the notations you made were wants and some of them were "must haves". In this exercise you must answer the following question:

Exercise:

"If the person had everything else I wanted except_____, would I be happy with this person?" Answer this in response to each characteristic that you noted in the first exercise above.

Circle each characteristic that you offer a "NO" response to. These are your MUST HAVE responses.

Obviously the MUST responses begin to limit the people with whom you are interested in sharing your life. Therefore you must be completely honest and true to yourself. It is not shallow to want someone who is good looking, or wealthy or even have a high degree of education. The question you want to answer for yourself is, is this REALLY a MUST HAVE. If someone has a history of sexual abuse or has committed murder those are most likely *"must NOT have" traits.*

You have now carefully discovered what you truly want and need in a love relationship. Your brain will screen in potential applicants. Similarly, as other people go through their day-to-day life, you will be screened in and out by them as well. This is a very important process to understand and appreciate. We cannot take "personally" that we do or do not fit into someone else's life plans. This process is one that most people go through on some level (conscious or unconscious and occasionally at the spiritual level). When two possible life partners meet each other, there is usually a "click" that happens shortly thereafter. This "click" is the "this is the one for me" response. The person may actually be only one of many, but when you feel the click, the experience is often one of "love at first sight".

EYE OF THE BEHOLDER

First Stage: Getting to Know You

The whole idea around dating is to get to know another person. Without this initial step, we would revert to arranged marriages and "mail-order brides". The first date can be tremendously exciting, or it can create extreme anxiety.

It could be that you have already gotten to know your date through association at work or in school. This makes it somewhat easier in that you already know that you have interests in common. Although having already spent time together under non-dating circumstances can make you more comfortable being with them, there may be some nervousness about how to turn an acquaintance into a romantic interest.

It may be embarrassing to admit that you have gained a new level of interest in that person. In these cases, go back to the chapter on flirting. Try out your flirting techniques and see if you receive the type of response that would indicate that the other party is showing enough interest in you that they would probably accept a date with you.

Another dilemma is when you have begun doing things with someone and you can't figure out if it is a date or a "do". Are you going out because the other person has a love interest in you, or are you going out as friends?

Indicators that your social activities are as friends rather than as a date:

> - *You are splitting the check.*
> - *There is no chemistry.*
> - *There is no flirting going on.*
> - *Not even a gradual move in the direction of closer intimacy is being indicated.*
> - *They are talking to you as though you are their Counselor.*
> - *They refer to their "ex" and other people they are dating.*

> - *They bring along work from the office where you work together, or suggest that you study together – and then really only study.*
> - *The energy between you feels more like siblings than lovers.*

When you know this is really a date:

> - *When they say that they want to take you out on a date.*
> - *The man expects to pay the bill most of the time.*
> - *When the electricity is tangible.*
> - *When there is distinct flirting between you.*
> - *You feel that you have gotten closer or more intimate by the end of the date – even if in small increments.*
> - *You have held hands or kissed, or have touched each other in some other tender way.*

Your choice of language can certainly help to clarify the situation. When you want to ask someone out, and it is meant to be a date, try one of the following phrases:

> *"I would like to buy you dinner."*
> *"May I buy you a drink?"*

> *"I am buying tickets to the theater,*
> *would you like to go?"*

If you would like to go out with someone as friends, and it is not meant to be a date, then you could say:

> *"Let's get together and have dinner."*
> *"Let's meet for a drink."*
> *"I'm going to a movie, would you like to join me?"*
> *"It's not a date, but a 'do'."*
> *"If you are in the mood for a drink and want company, call me."*

Believe it or not, the money issue can be very awkward for a woman if it is not clearly defined at the invitation. Many women will feel the need to pay their own way, or offer to pay. That is perfectly acceptable, however, it makes things go much more smoothly if it is obvious from the way the invitation was stated.

As awkward as a first date may end up being, it is an important step in the future of your relationship. Without a first date, you would jump from being acquaintances to being married. So we have no way around it, but to get used to the idea and practice getting good at it.

The first date is so important because it could really make or break the future of the relationship. But keep in mind that the decision concerning whether there is a future to this relationship is up to both of you.

Remember also that your initial contact with a person will set the tone for the relationship to follow. If you give over your power, or are willing to go to extreme lengths for the other person, then that might be what will be expected in the future. Be yourself, maintain healthy boundaries for your body, your energy, your money and your emotions. Only present to them what you are willing and able to deliver.

Too often we go into a first date feeling that the future of the relationship is all up to the other person. We turn that power over to them. In actuality, it is just as important that we remain confident, relaxed and poised. We want to be attractive and not rejected, but we, too, need to measure whether they are the right person for us to bring into our lives.

In some ways we have the opportunity to be either the one that chooses, or the one that is chosen. If we are choosers, we look for what we want and go after it. If we are among the chosen, we wait until someone shows interest in us and initiates the contact. There are advantages and disadvantages to both.

The advantage of being a chooser is that you are more in control of the direction of your life and are making the determination of what you want to draw into your social environment. The disadvantage is that there may be more opportunities for rejection.

The advantage of being among the chosen is that you know the other person is attracted to you because they approached you. The disadvantage is that you have to wait around, hoping that they will make the first move. In this position you have to be certain that you are always looking your best and groomed in a way that you will attract exactly the type of person you want. And then hope they are assertive enough to make the first move.

It is a good idea to find a comfort zone somewhere in between the two. Where you are attracting interested people into your life, while not too frightened to assert yourself in going after those who interest you.

As relationships develop, it is appropriate that the parties contact each other alternatively, with the man calling the woman about twice as often as she calls him. Although the man stays the main initiator, the energy should be flowing both ways. The woman needs to demonstrate care and attention, and indicate that she is as interested in the man as he is in her.

The first date is an opportunity for us to get to know about that other person. What are their likes and dislikes, their interests and hobbies, their style and personality? Do we think alike or do we have gaps in our thinking that no bridge can span? Do our personalities fit together smoothly, or is there unbearable friction?

It is perfectly acceptable, in fact rather important, that you are able to disagree with a point that is being made. It establishes that you have your own thoughts and ideas, that you stand firmly on your convictions and it makes you more interesting than just a "yes woman". However, avoid arguments or contrary attitudes just for the sake of individualism or challenge.

We are all individuals and we have our differences. And if it is decided that the two of you are incompatible, it doesn't mean that there is anything wrong with either of you. Not everyone will be perfect for you. Likewise, you cannot possibly expect to be the perfect match for everyone else. Don't take it personally. Just be kind, and move on.

In this first stage of dating, it is important to keep the options open. Don't place all your emotional eggs in one basket. Don't allow your psychological balance to rest on the reaction or compatibility prospects of one person.

Stay detached from the outcome until you have both gotten to know each other better. Date other people during this stage as well. This will keep your options open, give you other activities to keep you busy, and keep you from being too available to any one person (rarity raises your perceived value).

Through meeting a variety of individuals, you have an extraordinary opportunity to get to know yourself better. Many different types will be attractive, for a variety of reasons. You may encounter people whose interests and knowledge open doors and ideas to you that you never dreamed of. Your horizons will expand and you may discover aspects of yourself that have been unknown to you all this time.

While dating a variety of individuals, you will also learn what you don't like in yourself and in others. Some people will simply bring out the worst in us. These encounters, too, will provide learning lessons in personal growth. We can discover what we need to work on in ourselves to make us be even better human beings.

We can also determine what it is in others that we really cannot bear to tolerate. As negative as that sounds, it is important that we learn this. It is not uncommon for people to think that they can rescue another person from their troubles. They will get involved with a relationship that is abusive, or involves drug and alcohol addictions. It may be a noble cause to try to save them, but people with these types of serious issues need a professional counselor. An intimate relationship is not the arena for helping another person out of these troubles

Eventually we learn there are things we just cannot tolerate that were less obviously irritating at first. They could range from differences in tidiness, the handling of money, communication or whether the other person has children.

We don't always know whether we can live with certain differences until we experience a taste of what it might be like.

> *Dating affords us opportunities to learn about our preferences before we are committed to a situation that is more difficult to retreat from.*

It seems to be the human tendency to think that we can change another human being. The truth is that we cannot. Only they can choose to change themselves. And even when they make that choice, it can, in some instances, be a very long and arduous journey. The old saying, "You can lead a horse to water, but you can't make him drink", is really true. If some trait that you sense in another person is intolerable, respect yourself and them enough to gracefully move away from that relationship. Everyone will learn their lessons in their own time.

So, how do we know when it is right?

Although the nervousness of a first date can somewhat hamper the flow of your true and wonderful personality, there are a few signs that you will notice when things are really going great.

The conversation will flow. There will be a natural rhythm to the dialogue, each person contributing and taking turns in the spotlight. You will enjoy listening to them tell you their tales, and you will know that they are paying attention to what you have to say.

The sparks will be there. The flirtation will be high, natural, and lighthearted. There will be more touching at the end of the date than there was in the beginning.

Let there be laughter. Easy, natural humor, laughter, giggles (from being tickled inside, not from nervousness), smiles and comic relief all add to the pleasure and ease of the date.

You will each make the other person the center of your attention for the duration of the date. You will find yourselves in an imaginary sphere of energy and focus. When a date is successful, the rest of the world fades a bit, and nothing seems as important as just being here now.

What can we do to make things go smoother?

Once again, revisit the Golden Rule. "Do unto others as you would have them do unto you." What is it that would make *you* feel more comfortable? What could they do for you that would make it go smoother for you?

When you answer those questions, you can turn the tables and do exactly that for them.

> - *Give them a warm smile.*
> - *Make eye contact. Looking at someone directly in the eye gives him or her the sense that you are open, honest, and paying attention to them.*

> - *Make them feel welcome by reaching out to them, pulling them closer into your personal space.*
> - *Compliment them on their appearance or their accomplishments.*
> - *Draw them into a conversation and then be interested in their responses.*
> - *Be kind, polite, attentive to their needs, gracious and chivalrous (for men).*
> - *Be relaxed with them and let them follow suit.*

There are so many things you can talk about when you are first getting to know one other. After all, the whole world is available for you to talk about because it is all uncharted territory between you. You know so little about this other person that there is everything to discover and explore! Have fun with it and be natural.

Let your conversation touch on things you enjoy and know about, but look for those subjects that you have in common to really discuss in depth. Ask questions and make comments. Again keep it lighthearted and spontaneous. This is an opportunity to get to know one another, not an interrogation.

Some subjects are better left for later.

During the first date avoid the following topics:

- *Avoid talking about past relationships. Knowing that a*

*person is divorced or that they have
been available for a couple of years
is sufficient. You don't want to go
into the details of previous
relationships that either of you
have had.*

- *Don't talk about how badly you
have been treated by another
person. This will lower your value
in the other person's eyes. It may
indicate that you did not value
yourself enough to prevent or get
away from such a situation. If you
are willing to settle for an abusive
relationship, what does that
indicate about your choices in
relationships? This in turn, is
subtly insulting to your date in
that you are indicating lower
standards for yourself.*

- *Don't discuss other people that you
are currently dating. It's fine if
they know that you are dating
others. That may even increase
your attractiveness in their eyes.
Just skip the details.*

- *Don't lie. If you feel you have to lie,
there is something in you that you
should be working on, or he is
really not the right guy for you. If
you are trying to entice your date
with tales that are not true, you
will be getting them to be attracted
to someone who is not you. That is
no compliment at all. And
eventually a relationship built on*

lies will crumble. Why even bother? Surround yourself with people who love you for who you are. If you don't think you are lovable as you are, then do something to change that.

- *Avoid talking about commitment. That is a very scary subject, especially at the onset of a relationship. Commitment, marriage, children and building a family are all subjects to refrain from.*
- *This is not the time to discuss your finances or theirs. Talk of fortunes and misfortunes may put off your date, and give the wrong impression. If you have other things in common, there will be time to explore personal finances at a later date.*
- *Avoid whining, or negative talk about life, yourself or others.*

There are so many subjects that can be discussed that help you to get to know another person. There are also many clues to help you become even more irresistibly attractive to your date or new acquaintance. Here are some ideas for you to try:

- *Ask about interests, hobbies and travel.*
- *Inquire about occupation and what they like best about what they do.*
- *Do they have pets?*

- *Where have they lived during their life, and which place did they like the best?*
- *Determine whether you have friends in common. Casual name-dropping is acceptable as it gives you credibility and shows your connections to the world. Please avoid rattling off a list like a pedigree or with the air of competitiveness and boasting.*
- *Ask about favorite music, cuisine, drinks, and dance style.*
- *Wear or carry some object that is unusual and interesting. This gives you and your date a possible topic of conversation to break the ice. Your conversation will be on an object and the stories behind it, rather than directly on you.*
- *Use their name during the conversation. This personalizes your speech and lets them know that you are really connecting with them.*
- *Be interested in them. Give them your full attention. Even if the waitress interrupts, keep your focus on your date as you also handle the details of ordering.*
- *Be an interesting person. Have outside interests and activities that make you fascinating to the other person. Be aware of current events, be excited about what you do,*

develop reasons why people should be interested in being with you.

- *Keep the spotlight on them. Even though you divulge bits of information about yourself, let them ask if they are interested. Meanwhile, keep your focus on learning about them. They will be flattered – everyone loves to talk about him or herself. Further, you already know everything there is to know about yourself. The person who asks the most questions has the advantage of having learned the most during the date.*

- *Determine how you can make a difference in their life. Leave them feeling that you have had an additive quality. This should be something simple in the beginning of a relationship. You aren't trying to set up dependency or become their therapist. It should be something in relation to the conversation. For instance, if they have been wondering what to give their mother for her birthday, you may share with them something clever that you did for your mother.*

Where Should We Go?

A first date should be short and sweet. Since you are at the initial stages of discovering more about each other, a long date may be more than you can handle. It may become uncomfortable if you find that you really don't have much in common, or that you have run out of things to talk about.

If the date goes well, then a shorter date will leave you both wanting more. You will part with a pleasant feeling, looking forward to more contact.

First dates can take place at a coffee shop, over lunch, or at a comfortable bar over a drink. If you have determined that you both have dogs, you could have a first date walking them in the park and getting an ice cream from a vendor. A sporting event may be acceptable if both parties really have an interest in the event.

Second Stage: We're Seeing Each Other

"Will I ever see you again?"

Avoid saying that with any seriousness in your voice. You may come across as insecure, and that perhaps you were unconscious of the energy that occurred between you. If you were paying attention to all the signals, you should have a fairly good idea of whether you will see this person again.

Instead, you might make the statement, "I hope I will see you again." Or, "I will look forward to doing this again sometime." A man could also ask, "May I call you again soon?"

Of course, if things have gone really well, you could seal the deal right then and ask for a next date before the end of the evening. The woman can do this if she has specific tickets or an event to go to, and wants the man to escort her. Otherwise, the man should be the one to suggest the next date.

If one of you asks for the second date and the other person is non-committal or declines, then you will know that the spark may have been there for you, but that you didn't quite light their fire.

Regardless of what happens, avoid being clingy or appearing desperate. That is never attractive and generally pushes the person further away from you.

Most people are looking for chemistry when searching for the perfect date – their potential soul mate and life partner. If the chemistry is there, you will both be interested. If not, it might be just as well to remain friends, or just move on. When the connection is there, you will feel it at three levels: in the body, in the mind and in the spirit.

If a couple is connected in the body, they will find each other physically attractive. Their body language will mirror each other. They will enjoy the way the other person moves and dresses, and find them sexually attractive.

When a couple is connected in mind, they will have naturally flowing conversations, rarely run out of things to talk about, be comfortable in silence together, and have similar education and thought patterns.

Connecting in spirit, a couple finds they naturally understand each other. They simply have a knowing between them, a telepathic connection. At first sight they may have had a sense they had already known each other previously.

If one of these connections is in place, you will make good friends. When two pieces are present, it will feel so close. But you may always think, "if only there was just something more." When all three of these connections are in place, you will know that you have found a truly rare and wonderful soul companion.

If the sparks are there, you will surely be anxious to continue to get to know this wonderful person. It is time to explore longer and more frequent dates, discovering the many facets of this fascinating individual.

Once you have begun to see each other on a more regular basis, the conversations may turn to more in-depth issues. You will have the opportunity to really explore what the other person is like.

There are some behaviors that will enhance your attractiveness in the eyes of your dating partner. In addition to those points mentioned in reference to the first date, keep in mind the following:

- *Be attentive to your date. Don't ignore them or start looking at or flirting with others while you are out together.*
- *Be aware of your tolerance to alcohol. Drinking too much only makes you look foolish and you will probably say or do things that you will surely regret later.*
- *Drive safely. Do not endanger your date with driving antics on the road, or engage in road rage. Your attention should be on your date and the enjoyable time that you are having together.*
- *Arrive on schedule. Being late, except for some unforeseen delay, is*

rude and disrespectful to your date. Not showing up at all is inexcusable and should take you out of the realm of any possible future date.

- *Stay positive. No one wants to spend their spare time listening to someone go on and on whining and complaining.*
- *Pace your overtures and advances in accordance with your date's comfort and acceptance of them. It makes it uncomfortable and tiresome to be continually fending off unwanted advances.*
- *Turn off the cell phone. Unless your date agreed to go out with you with the knowledge that you were waiting for an important business call or an emergency call from your family, it is not impressive for your date to sit through a series of personal calls, or have the evening continuously interrupted by beepers.*
- *Just a dab of perfume or cologne is sufficient. Be aware that some people have sensitivities to chemicals and certain scents. A cloud of fragrance is not as alluring as a faint hint of sweet or spice as you come closer....and closer....*
- *The woman should ask the man out once in a while and pay the bill. Perhaps buy tickets to a special*

event or make a special dinner for the two of you.

- *Chivalrous behavior in a man is absolutely enchanting to most women. It will get you a lot further than most anything else that you could do. If the woman is offended that you want to open her doors, you can respect her wishes.*
- *Acting like a lady is probably one of the main reasons why a man wants to date a woman. If he were interested in being with someone that acts like one of the guys, he would go out with his buddies.*
- *Never ask the other person how they feel about you or declare your feelings prematurely. Their level of interest should be obvious to you by their body language, behavior and words.*

Just because you are dating, or have been for a while, avoid falling into the trap of having to know "where this relationship is going". Keep in mind that any relationship is only "what it is". Even if a person declares their undying love and presents you with a ring, when you have only been on a couple of dates, you are still in the formative stage of a relationship. No words will change the reality of the situation.

Furthermore, getting someone to declare their intentions does not mean that they will follow through. Just because they are ready to marry you in the throes of infatuation, when the excitement settles down, they may have no interest in rushing into a commitment.

It is not unreasonable to discuss with your dating partner what intentions they have for themselves. Are they interested in being in a committed relationship or are they interested in exploring being single for a long time? Do they someday want a family or are they averse to having children? That line of inquiry is legitimate. However, asking them if they intend on marrying you, or being the mother or father of your children, is not appropriate or relevant.

Remember that even at this point in the relationship, you are still not committed to each other. Until you have both come to a serious agreement on commitment, and that should include exchanging rings, you are still a free agent – and so are they. If things are not going well, either one of you is free to move on.

It is quite acceptable, and perhaps even desirable, for you to still be dating others. It is an exciting time of your life, full of people and adventure, but it also takes a certain finesse, so that you don't just drive them all away in exasperation. There are a few guidelines that will attempt to keep you in balance and out of trouble.

Once you have decided to continue dating someone, it is time to take a look at each other's lifestyles. If you are a "neat freak" and the other person lives as though a cyclone just passed through, there will be some fundamental problems down the road should you choose to combine your lives.

As silly as any one of these items may seem, in the long run, and especially in combination, they could destroy a long-term relationship. So during this early stage of discovering the possibilities of compatibility, look for these things:

- *Neatness vs. sloppy*
- *Financial stability and how they handle their checkbooks*

- *Spending habits – are they too frugal or out of control?*
- *Intelligence – are you at similar levels or is there a disparity?*
- *Social status comfort zone – Do you camp and they prefer the Ritz?*
- *Introvert/extrovert – Will they want to stay home and read while you want to party all night long?*
- *Giver vs. Taker – Does one party take advantage of the other in any way?*
- *Children/no children – Do you both agree to have or not have children?*
- *Pets – Will you get along with their pets or be allergic to them? Will their pets come before you and the relationship?*
- *Can you tolerate their family or will it be the Montague's and Capulets?*
- *Smoker vs. non-smoker*
- *Do they have habits that are out of control? Drinking, drugs, etc.*
- *Monogamous? Polygamous? Able to make a commitment?*
- *Any other skeletons in the closet – criminal records, violent past lovers who are lurking around, diseases (sexually transmitted and otherwise), psychological problems, and so forth.*

Assuming that things are still looking pretty optimistic, it is time to meet their parents and their friends. This will give you further information about the person, and the opportunity to be together in a group setting. How do they introduce you? Do they continue to give you the same attention and respect as when you are alone?

They will have a similar opportunity to meet your friends and family. How do you feel about them when they are in your personal circle of friends? Are you proud to be with them? Do they make you feel good? Will they integrate into your lifestyle with ease?

Stronger discussions of values and beliefs can be broached at this point as well. If serious discrepancies are noted at this point, you still have the opportunity to move on to other dating choices. By now you may have learned a lot about yourself, making you more inclined to commit, or giving you the realization that you are not yet ready.

At this point you will have learned what you can live with....and what you know you can't live without!

Developing a pet name for someone is a further means of flirting at the stage when the relationship has begun to develop. A pet name should be unique and personalized. Perhaps it will remind each of you of a special moment or endearing trait of that person.

Using "Honey" and "Sweetheart" are not particularly individualized. A friend of mine once told me (ML) that he called all of the women he had relationships with "Honey" because that way he would be sure that he didn't call the present one by the name of a previous lover. That has certainly tainted my view of being called "Honey"!

Although the dating process seems to be filled with rules, it can also be some of the most enjoyable and exciting times of your life. Set your sights high that you will meet the right person for you. Don't settle for less than you deserve. And always be prepared to present to the world your most irresistibly attractive self.

Stage Three: We're Getting Serious

If you have come through the entire dating process and find that you both want to spend the rest of your lives with each other, you are a fortunate couple, indeed. Being in a committed, sincere and loving relationship can be the most rewarding and fulfilling part of a person's life.

> *In a committed relationship, as in the early dating stages, it is important to maintain our best efforts at being attractive.*

Remember that it isn't your job to change another human being. You are not in a relationship for the purpose of changing them. The only proper reason to be there is because you share the same goals, desires, lifestyles and are madly in love with the person who they are.

One evening I (ML) was enjoying the company of several of my girlfriends and our conversation turned to the topic of relationships. One wise friend piped up and said, "When two people marry, the woman is planning all the ways that she can change her partner, while the man goes into the relationship hoping she will never change a bit from who she is on their wedding day." Women are more typically the ones that continually look for ways of self-improvement and change, while men typically just want everything to stay the way it is. This can add to the never-ending drama that occurs between Mars and Venus!

Just as you want to wake up every morning and fall in love all over again with that special someone, you will want him or her to do the same. Flirting, going out of your way to please them and make them happy will be as rewarding to you as it is pleasant for them. When you give another person your love and attention, it will be so easy for them to be open to you and give back the same good energy.

- *Keep the door to communication always open.*
- *Update your look and stay stylish*
- *Exercise and watch your diet. Don't let your body fall apart.*
- *Maintain your health, grooming and good habits.*
- *Listen to your partner and hear even what they are not saying to you*
- *Be sensitive to the other person's needs.*
- *Stay active and youthful in your body and in your thoughts.*
- *Be supportive and encouraging to your partner.*

- *Allow them to be vulnerable, providing a safe space in this harsh world.*
- *Compliment your partner frequently.*
- *Show appreciation for what they do and who they are.*
- *Always present a pleasant and loving picture of your partner and your relationship to the public.*
- *Above all, keep the passion high between you.*

Over time, in a committed relationship, it can be easy to let little things build into big things that work to deteriorate the quality of the relationship, leading, perhaps, to a break up. It is important in any relationship to keep the communication flowing. It is vital that both partners participate in the strengthening of the bonds, if the relationship is to last.

EIGHT

Personal Magnetism In Intimate Relationships: Can Women and Men Really Talk?

Women are the relationship-oriented communicators of our species. They want to talk with men. They want men to listen to them. And when they find men who listen, they are magnetized to the shockingly evolved male. (At least that's what women tell us!) Men are typically task and results-oriented people. Women tend to focus on developing relationships and intimate communication with those they care about.

For millennia, men have been dumbfounded by the ways of women and vice versa. The reason, of course, is simple. Men and women do not think alike. If a woman were to approach a man at a bar and ask the man out for a date without previous communication, the man would almost certainly say, "yes", if he found the woman attractive. This is the way men are. Men are almost honored if they believe that a woman would find them so physically attractive as to ask them out without even knowing what they are like as a person!

Women, on the other hand, typically find such a request insulting! Women need to feel that a man is interested in them as a person before allowing a positive response by accepting a date. Most women can justify accepting a request for a date if they believe they have earned the respect of the man requesting their presence.

Women prefer to participate in communication where an equal amount of disclosure takes place between them and the other person. When people engage women in communication, those people who share somewhat intimate experiences about themselves tend to be appreciated more.

We teach a class called "Irresistible Attraction" about once per month for various adult education facilities all across the United States. (The class is what gave us the idea to write this book, by the way!) It is a one-night experience that allows men and women to learn what it is that members of the opposite sex find attractive about them. Women regularly tell me that they want to develop a relationship of some kind with a man before he asks them out. This relationship (nothing more than a conversation, really) could be as short in duration as one hour! Women simply want to be taken seriously. Women know that men do not develop relationships with their buddies just by looking at them and inviting them over to watch a football game. Women want to be treated on a similar level that allows her to feel as if the man knows her to some degree before attempting to build a more intimate and deep relationship.

Regardless of your gender, when you are dealing with women, remember that women want mutual disclosure in communication and that is the normal way of developing a deeper relationship with them.

Give a Man a Problem to Solve

Men are problem-solving creatures. (We are other kinds of creatures too, but we are genetically wired to solve problems.) When people talk to men, men go into problem solving mode. If you want to get a man's attention, then you want to present to him a problem that he can solve. Most women do not understand this male compulsion to solve problems. Women prefer to discuss their problems and share their experiences with no expectation of solutions. In fact, women often feel cheated when a man rapidly offers a solution to a problem or difficulty. With men, the opposite is the case.

Present a man with a challenge and he will communicate with you. He will ask you questions to discover the cause of the problem. He will search with you for the spark that started the fire and then he will put it out...and quietly go back into hibernation. Most men are simple creatures, driven by their libido and distracted by their left-brain when the desires of their libido are not being met.

Men listen when they believe they can help solve a problem. Men listen when they think what they hear will help them solve a future problem. Men listen when what they hear will in some way make them more productive in some way, for some purpose. Men are collectors of problem-solving tools. Many men have garages filled with screwdrivers, hammers and saws. Some men have huge libraries of books that have the answers to the world's problems. Still other men have guns, fishing poles, bows and arrows, and knives to solve the problem of feeding the family. If you think of men as being problem-solvers, you will be a step-up on gaining the attention of the one man you want to talk to!

Most men need a self-esteem boost, but you would never know it from looking at them or listening to them. They appear quiet and confident when they are really just being quiet. They *do* tend to solve problems well. Men feel greater self-esteem when they are taking charge of a situation and meeting a challenge that they can handle. If you want a man to listen to what you have to say, help raise his SEQ (self-esteem quotient)!

You could help a man raise his self-esteem by telling him he is wonderful, good looking or kind. (Men always appreciate these flattering tidbits, even when they aren't true.) If you want to communicate with a man, then engage him in activity where his self-esteem will be enhanced. Work with him on a project that he is good at.

Have you ever noticed a group of men surrounding an automobile with an open hood and talking away? There is a problem to be solved, and communication comes easy. The man who has his "hood open" (whether he is the hunter, the carpenter, the scientist, the artist) is primed for listening to relevant conversation. He will share his wisdom and listen to reasonable discussion, however tangential.

> *If you want a man to listen to you, ask him to help you solve a problem.*

In the balance of this chapter you will learn the most common mistakes people make in interpersonal communication. Avoid these mistakes and you will find people magnetized to you!

Mistake #1

One-Upping Your Friends:

How to Invalidate the Experience of Others

Mary: I had such a bad day today. I got stuck in traffic for over an hour.
John: That's nothing. I was stuck for two hours.

John just invalidated Mary's experience and cut off rapport and communication, didn't he? How could John have handled this so that Mary would know that she is important to John as a person and at the same time share his challenging day?

Attraction Tip #1

Listen Fully Before Sharing a Similar Story

Mary: I had such a bad day today. I got stuck in traffic for over an hour.
John: That's terrible. You know they really need to build another lane on the major highways around town. Were you in a hurry to get home or was it just that you got annoyed because of the pokiness of traffic?
Mary: Neither. I had a meeting I had to get to and I was late so the boss thought that I was putting a low priority on the meeting when really I was racing to get there.
John: Were you able to explain it to him?
Mary: No. There were too many other people around and excuses are excuses to the boss.
John: Is it going to affect your job?

Mary: Nah, my sales are in the top ten percent, I just don't like looking bad in front of the whole group. Not really that big a deal.

John: I understand. Today I was heading home, just for dinner, no meeting or anything, just wanted to get home to Christa and the kids and there was an accident on the off ramp. Traffic was backed up forever...but at least I didn't miss a meeting.

In the first scenario, John invalidates Mary's experience in a manner that is typical and common in our culture. This happens every day and we all have offended others at one time or another. What we want to do to improve our relationships with others is to get out of ourselves when talking with those we care about, and listen to the concerns and needs of others.

It is almost always best to let someone talk it out or talk out their problem to the point where it no longer bothers them before we begin to express our needs and interests. The reason is simple. When a person is wrapped up in their own world (what I call internalized) they are going to be almost completely disinterested in what is going on in your world. When a person has been listened to and is able to get it all out, then the person who was in their own world (internalized) will be able to slowly get out into the real world and feel empathy and concern for others, in this case, you.

The common mistake we all make in communication is trying to communicate two or more messages in a conversation at the same time. This takes away from mutual understanding and a true concern for the interests of both people. When people let one person disclose all of their feelings and frustrations about some difficulty, that person is then far more likely to listen and then care about the difficulties and frustrations you face.

Mistake #2:

I Don't Want to Listen to You.
Let Me Just Solve Your Problem.

These types of problems in communication are often magnified between men and women.

Women often feel the need to express their problems without the absolute necessity of having them solved by someone. They simply want people to listen and empathize. Here are a couple of examples of how men often don't understand women's communication needs.

Jan: Today was such a bad day at work. The boss yelled at me for not getting carbon copies to all the people at the meeting.
Richard: No problem. Next time just print out extra copies so you have more than enough for everyone.
Jan: You don't have to tell me the obvious Richard; I'm not an idiot.
Richard: I didn't say you were an idiot. I just said it would be good to have extra copies of all the reports. If you would just think ahead you wouldn't have these problems.
Jan: I do think ahead, Richard. I didn't know that all of the Assistant Managers would be there, too. Why do you find it necessary to make me feel stupid all the time?

Richard: I'm not trying to make you feel stupid. You could just ask the boss how many people are going to be at the meeting and then prepare for them and a few more.

Jan: I do check the roster so I know how many people are coming. There was just a miscommunication as to who else would be there. It wasn't like it was my fault.

Richard: Whatever. I better go cut the lawn. (Leaves)

Jan: (Sits, depressed)

Attraction Tip #2

Sometimes Listening is Better than Solving Their Problem

Richard and Jan have just had the beginnings of what will likely become a huge argument and it started over next to nothing. How could Richard have saved the day for his frustrated wife? Listen to the better flow of communication below as Richard realizes the need for Jan to simply share the challenges of her day and not have them solved.

Jan: Today was such a bad day at work. The boss yelled at me for not getting carbon copies to all the people at the meeting.

Richard: What happened?

Jan: Well the stupid General Manager didn't tell me that all the Assistant Managers were also going to be there so I ended up looking stupid because I didn't have reports for everyone at the

meeting. I was so ticked off, I could have just cried.
Richard: *Did anyone say anything?*
Jan: *Well, there just weren't enough reports for everyone and I felt terrible. It's not like anyone said, 'Oh Jan, you are such an idiot.' It was just that I looked bad. I even called ahead of time to see how many reports I was supposed to bring. I was just so embarrassed.*
Richard: *Must've really bummed you out. I'm sure sorry to hear about how unaware some of these people are that you have to deal with.*
Jan: *Yeah, they expect you to be God and you just aren't.*
Richard: *Well, I don't know about that, you're pretty close in my eyes (a bit of a grin).*
Jan: *(a bit of grin and with a playful tone in her voice) Oh you...*

Happy endings are almost always very easy to accomplish, especially when we realize that we can communicate with people in the fashion that they want and expect to be communicated with. Many women have a common trait that they do not want men to solve their problems, they want men to listen to their problems and empathize with them. For men, this concept is foreign. Men are problem-solving creatures by nature. Watch the following scene where Richard really needs help in solving a problem, but this time Jan responds as she would if she were communicating with another woman.

Mistake #3:

Not Helping Men Solve (some of...) Their Problems

Richard: The weeds are ridiculously thick in the lawn, the lawn needs to be cut, it needs to be watered and I just don't have time to do this and the painting, too.

Jan: Honey, I sure am sorry you have so much work to do. I really understand.

Richard: Understanding is great but it doesn't paint the house and water the lawn. Can't you at least help me in some way?

Jan: Honey, I know you are frustrated.

Richard: Of course I'm frustrated, there are 30 hours of work to do today and I only have 24 to get them done. I'm going nuts.

Jan: Well, pretty soon it will be winter and there will be less outdoor projects to do.

Richard: That is so typical. Can't you for once just say, 'Hey honey, howzabout I water the lawn for you?'

Jan: I just don't understand you. You are always saying I'm lazy. I work my backside off around here.

Richard: Well, what the heck do you think I do for you?

Jan: I think that I don't complain half as much as you do and at least I listen when you have problems. You just go off telling me what to do all the time.

How could this soon-to-be Titanic-like disaster have been avoided? Let's look at the new scene once Jan understands how Richard, and indeed most men, need to be communicated with when they are under stress in trying to solve problems.

Attraction Tip #3

Gently Help Men Solve their Problems

Richard: The weeds are ridiculously thick in the lawn, the lawn needs to be cut, it needs to be watered and I just don't have time to do this and the painting, too.

Jan: You're right. That is too much for one person to do. We can do a couple of things. I can water for you after you cut the lawn or we can pay one of the kids on the block $20 to cut the lawn and then water. What do you think we should do?

Richard: I don't really know. I hate to pay the money, but I really need a break sometime this weekend.

Jan: Then forget the money, I'll call the neighbor boy and we will get the lawn cut and watered and that way you can get to paint the house without feeling quite as overwhelmed.

Richard: Thanks, honey. Maybe there will actually be time for us sometime today.

Men and women's communication needs are often very different, though they have the same ultimate purpose, which is to experience some sort of happiness or pleasure. The road to happiness and pleasure is paved with talking to people the way they need to be communicated with and not the way *we* want to be communicated with. When in doubt, the rule of thumb is that women want people to listen to their problems. They want empathy and understanding and an appreciation for the experience they are having. They are not looking for solutions to their problems unless they come forward and ask for assistance in seeking solutions.

Men, on the other hand, tend to communicate with the intent of solving problems and participating in goal directed behavior. They prefer not to participate in communication that "serves no purpose." They often perceive this as a waste of time.

The exception to this rule happens when men get lost driving, which happens with some regularity. Men do not want assistance in finding their way to and from destinations, so when you find yourself lost with a man, keep it to yourself and act as though little, if anything, has gone wrong!

Mistake #4:

Ignoring the Values and Beliefs of Your Listener

Miscommunication often happens when we do not consider the personality, the needs, the ideals, the values or beliefs of those we are communicating with. Discover the make up of a person's personality, her beliefs, her values, her ideals, her goals, her wants and interests, and communication becomes like rowing a boat downstream...very simplified.

Attraction Tip #4:

Learn to Tolerate Most Values and Beliefs of Others

Once we understand what makes another person tick, we can easily build rapport and enthusiasm within our communication with them. Learning the true values that others hold dear to them is like having the keys to their soul. Once you have experienced someone sharing their deepest beliefs and values, respect and honor those values.

Mistake #5:

Arguing With People Who are Obviously Wrong

Jackie: I hate parents who let their kids stay up until 10 PM.

Jane: What a stupid thing to say. Where did you ever dream that up?

Jackie: You always keep your kids up late and have no time for yourselves.

Jane: That doesn't mean I'm stupid, Jackie. You aren't exactly a perfect mother yourself, you know.

Jackie: You don't have any rules. You let your kids run the house and you'll pay the price later.

Jane: You're just being stupid. We have lots of rules. We simply don't worry about what time our kids go to bed.

Jackie: Well, maybe you should.

The scene is a familiar one among friends with children. Both of these women have different values about bedtimes and apparently rules in families, as well. It's also obvious these two women are close enough to express their opinions with each other. However, they are also at a point in their relationship where the words they use are often cutting and meant to hurt...just a little. Here is how words can be transformed from weapons to building blocks for better relationships.

Attraction Tip #5

Hear People Out Who have Opinions That Are Obviously Wrong

Jackie: I hate parents who let their kids stay up until 10 PM.

Jane: What do you mean, you hate parents who let their kids stay up until 10 p.m?

Jackie: It's sinful the way some parents don't have adult time. I've always said that there is adult time and there is children's time.

Jane: Huh, never thought of that. Is that how your parents were or did you figure out this just worked for you guys?

Jackie: When I was a kid, we were told that children were meant to be seen and not heard. We weren't allowed to be up after eight o clock on school nights and if we were, we would get our backsides paddled.

Jane: How did you feel about that as a kid?

Jackie: I didn't like it then, but I can see the wisdom of my parents. We learned discipline at an early age.

Jane: What do you think the best part about this is for the kids?

Jackie: Jane, they learn that there is adult time and there is children time. It's important for Mom and Dad to have time together, too.
Jane: Is that the most important part about late night time together for you guys?
Jackie: You bet.
Jane: Then I think you guys are doing it right. What makes me wonder is whether or not that is true for all families and circumstances. One thing is for sure is that having some adult time would be mighty nice sometimes.

Jackie has decided and has been "programmed" to believe that staying up until 10 PM for children is "sinful." She's obviously wrong. There are instances when it is good and, no doubt, instances when it is bad from family to family and circumstance to circumstance.

Notice how Jane defuses the potentially volatile conversation into a non-issue. Whether a family chooses to have adult time or not is really very little reason to attack your friends. Meanwhile, there is no reason for Jane to argue or fight about an issue which Jackie is being insensitive. In the future, there will come a time when Jane will want to share precisely how insensitive Jackie was, in a very clear manner. However, this is not the appropriate time. This is the time for de-fusing and not for enhancing a non-issue into an issue.

We all speak insensitively on occasion. Generally we speak insensitively when we have been dealt with harshly on the same issue ourselves in the past. If we tend to be overly strict as parents in some circumstances, these are often areas that we had a great deal of resistance to when we were young. When you hear someone making bold and obviously foolish statements, don't ever try and make them see the light of day. They won't when their fuse is lit.

When people cling to strong beliefs, regardless of how accurate they are, they will tend to defend these beliefs forever, until death do they, and their belief, part.

Mistake #6:

Tell Him What to Do With His Money

Consider the following scenario where a bad financial decision is made because of a foolish belief.

John: Hey, with that big lump sum you just got from your bonus check are you going to pay off that 18% credit card you've been complaining about for the last few years?

Richard: Heck no. We'll just charge 'em up again like we did last time we paid 'em off and then all that work has gone for nothing. I'm going to stick the money in the bank and keep it safe.

John: But, you'll only get 2% in a bank. You will basically get an 18% return on your money if you pay your credit cards. Think about it, that's a lot of money!

Richard: Well John, when you get your bonus check, you pay off your credit cards. We are not going to charge anymore on our cards. They're full, so we can't charge anymore. I'm gonna keep it that way.

John: But Richard, that makes no sense. You'll save thousands of dollars if you wipe out the credit card debt!

Richard: But in six months we'll have it all back again just like last time. Forget about it, John.

Clearly John is in the right. Paying off the credit card(s) is a far superior option and a substantially better return than the pittance he will get from a bank. The belief he has about charging up his credit cards is legitimate. What has happened in Richard's mind is that he associates charging up the cards again with paying off the balance and he doesn't want the cards charged up again. So the question is, how can John advise his less financially astute friend without pushing his hot button of keeping the credit card open for abuse?

Attraction Tip #6

Use The "They Technique": Direct People to See the Light for Themselves Without Hurting Their Feelings

---

John: Congratulations on getting the bonus.
Richard: Thanks, I really worked my back side off to get it this year.
John: Any plans for my newly wealthy neighbor?
Richard: Gonna save it!
John: Smart man. A lot of people would go blow it on toys or something.
Richard: Yeah we've done enough of toy buying for a lifetime. It's going to the bank.
John: Good for you. Hey whatever happened to that monster credit card debt you had amassed?

Richard: *Still got it and that will never happen again either.*

John: *Ya' know Fred's wife tore up all their cards except for one, which they both decided to keep for emergencies. They had amassed about 10 thousand dollars, just like you guys, and then they ripped up the cards. Since then, they're still paying 18% per year which is like having $2,000 sucked out of their wallets every year but at least they aren't digging themselves deeper in debt. Can you imagine? $2,000 per year in interest is $10,000 in five years! In other words, if they pay the minimum payment only, their credit card debt will never go down and they have to work a whole month just to pay the bank. Amazing isn't it?*

Richard: *How do you figure they work a whole month to pay off interest?*

John: *Well, they probably make about $4,000 a month between them right?*

Richard: *Yeah, sounds right.*

John: *Well, if you make $4,000 you have to pay Uncle Sam and the state about $1,000 of that. That leaves you with $3,000 right?*

Richard: *Sure.*

John: *Well their minimum payment will be about $250 per month. Maybe a little more. That's another $250 per month all going to interest. That's $3,000 per year so they have to work an entire month or maybe a little more just to pay the credit card. Revolting as it seems they each go to work every day for a month, just to pay the credit card. Too bad they didn't have a lump sum like you do to get out of that prison.*

Richard: Huh....never thought of it like that...but if you pay it off you just tend to charge it back up again.

John: Yeah, they used to be pretty irresponsible with their money, but they made a deal with each other to simply never do it again...and Fred says they haven't. They're going to be hurtin' for awhile, for sure, but in the long run, they'll get out of debtor's prison.

Richard: Maybe we'll get out of jail now...I gotta talk to Janet and see what she thinks.

John: Good thinkin'. Yeah, now that you mention it, you'd save a fortune and wouldn't have to work for Uncle Sam AND the credit card anymore...just Uncle Sam. (grins)

Richard: Yeah...

John used a technique that I call the "They Technique." This technique is used when you can cite someone else as an example of good or bad behavior. Then you gently imply the solution to the problem, without stating to the person you are talking to how obvious it would be for him to do likewise. You never directly assert that Richard should or shouldn't do something. You simply tell a story without even saying that this is a good solution for Richard as well. It's just a "story".

Using the "They Technique" allows you to help your friends and family members in a manner that lets them decide for themselves if they might have a better choice in a current situation. When people are participating in behavior that is obviously wrong or foolish, and you want to lead them to a solution that is definitely in their best interest, you can use this technique to help them get what they really want and deserve in life.

Mistake #7:

Fueling the Fire for a Potentially Explosive Subject

Anne: You know I saw this sign on the road for Planned Parenthood and I thought to myself that is nothing more than planned death. What a disgusting organization.

Chris: Anne, you will never learn. A woman has a right to choose. Why don't you just go blow up all the clinics in the state and become a martyr for your anti-abortion stand?

Anne: No, Chris. Not anti-abortion. Pro Life.

Chris: You want women to die in childbirth and couldn't care less how a woman feels when she's been raped and would be forced to carry some criminal's baby. You have no feelings whatsoever toward other people.

Anne: YOU are the one who doesn't care about feelings. Think of that little baby. Who makes her choices? You make me sick.

Few issues are more polarizing than that of abortion. Anne and Chris have been girlfriends for many years. They take their kids to the park together, they play tennis together, but they disagree on this volatile issue that has no societal right/wrong. It is very personal and about half of all Americans fall on either side of the debate. How could Chris have handled Anne's original outburst?

Attraction Tip #7

**Discover How People Came to Believe
in Their "Bomb Issues"
(Such as Religion and Politics)**

Anne: *You know I saw this sign on the road for Planned Parenthood and I thought to myself that is nothing more than planned death. What a disgusting organization.*

Chris: *Anne, how did you come to think about Planned Parenthood as a disgusting organization?*

Anne: *Because they counsel innocent women to become murderers of their children.*

Chris: *I know how strongly you feel about the Pro Life view and I really respect your feelings about the life of children, especially those that are unborn. In your mind, what they sometimes do may mirror what you just said. Sometimes though, they help people prevent pregnancy in the first place which eliminates the choice of having an abortion or not, and I think we both agree that is good, true?*

Anne: *Well yes, that is true. We both agree that preventing unwanted pregnancy in women is a good thing. It's what happens after pregnancy that always divides us.*

Chris: *You're right it does, but I want you to know that I respect your point of view and I know that it is made with sincere love in your heart and I hope someday you'll feel the same about me and my beliefs. Until then, we'll just have to agree on helping women use birth control*

effectively   so   there   are   fewer   unwanted pregnancies, period."
Anne: *I still think I disagree...but OK.*

Anne has been successfully defused and Chris, through her responsible and insightful communication style, has de-fused herself, too. No doubt, Chris can debate her own views with violence in her voice as well. However, instead of creating ill will where none needs to exist, she lets Anne vent her feelings then finds a point of agreement. (Face it; these women aren't likely to personally experience an unwanted pregnancy in their life. Why create resentment between these two women?)

There is rarely a need for people to make each other wrong and be right themselves. We all have the desire to be right. Over the years we have all been "made wrong" by others many times. Some of us have been made wrong so many times that we feel the need to not only be right but to make sure others know it. Some people even have the need to be right so much that they make others "wrong". These "victories" slowly erode the quality of relationships and make for difficult communication. When a person begins to attack with their language, the best defense is often to simply let the person say their piece and continue on with another topic or issue.

When your beliefs come into conflict with another person's beliefs, there probably is little purpose in attempting to make the other person wrong and you right. Simply let the person share their belief with you. Try and understand how that belief was created. What experiences led them to this belief? Learn what creates beliefs in the people you talk with. Be gentle. Be respectful. Be curious and be tolerant. These are all traits of good communicators and people who communicate with love.

Emotions are at the core of any human and their behavior. As you watch people act and react every day, you will notice certain patterns of behavior. People tend to display the same emotional responses week after week, month after month, and year after year. Emotions are a part of people that do not readily change.

If you find that using these tips don't help you move past the areas in which you and your partner don't see eye to eye, it may be time to look to someone else. For most people however, these tips will be enough to make you look like a brilliant communicator and caring individual!

Epilogue

We encourage you to practice the techniques and visualizations that have been presented in this book. Only with practice will it become totally natural to always show the world your most irresistibly attractive self.

Some people say that they only want to be natural, that to work at being attractive is phony or just not them. However, what we deem to be natural is just the habits that we have acquired over the years. If our habits are to be unkempt or loud and boisterous, they are just that. Habits. Habits can be changed.

And let me ask you this: How is it working for you so far? Are the communication skills that you have always used been effective? Are your personal relationships all that they can be? Is your physical appearance attracting the numbers and types of individuals that you would like to attract?

If you answered, "No" to any of the above questions, then perhaps acting natural is not netting you the results that you would like. In mathematics, we say that if you keep coming up with the wrong answer, you need to change one of the factors in the equation.

So this book has given you some choices of action that will help you to achieve the outcome that you seek.

Not every technique will be right for you. And not every person will want to achieve the same results. However, with the tools that have been presented here, you will have the opportunity to redesign your life, open the pathways of communication with others, and improve your ability to always be irresistibly attractive!

Although some of these suggestions appear to be tedious, or difficult to master, the important thing is to remember to have fun along the way. You have spent (how many?) years using techniques that have not always given you the results you were seeking. What have you got to lose by trying something a little different?

Once you get started making steady improvements and changes, you may see that the attention and results that you get are so exciting that you find yourself looking forward to trying even more new ideas. Each change that you make will have a ripple effect throughout every other part of your life. As you become more attractive to members of the opposite sex, you will find that your energy is more youthful and energetic, improving your work performance and how you are looked upon in your career. And so on, and so forth. Continually changing and improving, making all areas of your life more enjoyable.

It is your life. Live it to the fullest.

And be prepared to enjoy being irresistibly attractive!

Bibliography

Brandon, Nathaniel, 1994. The Six Pillars of Self-Esteem, Bantam Books, N.Y.

Buss, David, 1994, The Evolution of Desire, Basic Books, N.Y.

Furman, Leah and Elina, 1999, The Everything Dating Book, Adams Media Co., Holbrook, MA.

Gray, John. Men are from Mars, Women are from Venus, 1992. Harper Collins, N.Y.

Hogan, Kevin L., 2000. Talk Your Way to the Top: Secrets of Communication that Will Change Your Life. Pelican Publishing, Gretna, LA.

Hogan, Kevin L., 1996. The Psychology of Persuasion: How to Persuade Others to Your Way of Thinking, Pelican Publishing, Gretna, LA.

Pease, Allan, 1981. Signals: How to Use Body Language for Power, Success, and Love. Bantam Books, N.Y.

Spiegel, Jill. 1998. Flirting with Spirituality. Goal Getters, Mpls.

Townsend, John, 1998. What Women Want – What Men Want: Why the Sexes Still See Love and Commitment so Differently. Oxford University Press, N.Y.

Wright, Robert. 1994. The Moral Animal: Why We Are the Way We Are: The New Science of Evolutionary Psychology. Pantheon Books, N.Y.

For More Information

Get your Free Copy of Kevin Hogan's 163-page e-book, **Mind Access**! Learn how to ask just two questions and know someone's buying profile! This is some of the finest material ever written about influence and persuasion!

And as a super bonus gift, get Kevin's 179 page e-book, **Breaking Through the 8 Barriers of Communication**. You'll discover the 8 ways to avoid turning people off and instead, build powerful LASTING rapport. A full length book!

And.... the gift that keeps giving, a one year subscription to *Coffee with Kevin Hogan,* the e-zine that everyone reads about influence, selling, body language and personal development every Monday morning. Go to http://www.kevinhogan.com/bonus.htm today!